TRACKS IN DEEP TIME

JERALD D. HARRIS AND ANDREW R. C. MILNER

TRACKS IN DEEP TIME

THE ST. GEORGE DINOSAUR DISCOVERY SITE AT JOHNSON FARM

THE UNIVERSITY OF UTAH PRESS

Salt Lake City

The Defiance House Man colophon is a registered trademark of the
University of Utah Press. It is based on a four-foot-tall Ancient
Puebloan pictograph (late PIII) near Glen Canyon, Utah.

19 18 17 16 15   1 2 3 4 5

LIBRARY OF CONGRESS CATALOGING-IN-PUBLICATION DATA
Harris, Jerald D., 1970-
  Tracks in deep time : the St. George dinosaur discovery site at Johnson Farm/
    Jerald D. Harris and Andrew R.C. Milner.
      pages cm
  Audience: Grade 9 to 12.
  Includes index.
    ISBN 978-1-60781-437-5 (pbk. : alk. paper)
    ISBN 978-1-60781-438-2 (ebook)
1. Dinosaur tracks—Utah—Saint George. 2. Dinosaurs—Utah—Saint George.
3. Paleontological excavations—Utah—Saint George. 4. Saint George (Utah)
I. Milner, Andrew R. C. II. Title.
QE861.6.T72H37 2015
567.909792′48—dc23
                                    2015015503

All life restorations of extinct plants and animals were illustrated by
H. Kyoht Luterman, and photographs of SGDS fossils were taken by
Anna Oakden, except as noted in the caption credits.

Printed and bound in Malaysia.

CONTENTS

FIGURES

ACKNOWLEDGMENTS

Since the initial discovery of the fossils at the St. George Dinosaur Discovery Site at Johnson Farm (SGDS), hundreds of people have participated in finding, documenting, preparing, researching, and teaching about its fossils. Special thanks go to Sheldon and LaVerna Johnson for their fantastic discovery and for their altruism, which ensured that these amazing and important fossils are preserved for the future. The added generous donations of fossils and efforts in the site's behalf by Darcy Stewart, Paul Jensen, Jeff Chapman, Matt Musgrave, and the Washington County School District have immeasurably increased the importance, value, and prominence of the site. The efforts of members of the DinosaurAH!Torium Foundation, the City of St. George Advisory Board, the Utah Geological Survey, the federal Bureau of Land Management (BLM), and the National Park Service (NPS) deserve a great deal of credit for helping the SGDS succeed and flourish. In particular, Utah state senator Wayne Harper, Utah representative Martin Stephens, US senator Orrin Hatch, US representatives James Hansen and Jim Matheson, and Utah governor Michael Leavitt made funding possible to establish and build the SGDS museum. St. George mayor Daniel McArthur, the St. George City Council, and Kent Perkins and Gary Sanders of the St. George Leisure Services Department facilitated construction of the museum building and its subsequent maintenance.

Many scientific advisers have been instrumental in interpreting the fossils and geology of the site, including Sid Ash, Bob Biek, Brent Breithaupt, Richard Bromley, Paul Bybee, Marjorie Chan, Karen Chin, Don DeBlieux, Rob Gaston, Gerard Gierliński, Janice Hayden, Martha Hayden, Andy Heckert, Adrian Hunt, Randy Irmis, Jim Kirkland, Allan Lerner, Martin Lockley, Fred Lohrengel, Spencer Lucas, Scott Madsen, Neffra Matthews, Paul Olsen, Bob Reynolds, Jennifer Reynolds, Michael Schudack, Leif Tapanila, Don Tidwell, Garrett Vice, and Grant Willis. The outstanding efforts of SGDS director Rusty Salmon and former coordinators Jon Christofferson, Anneli Segura, and Theresa Walker have been critical to ensuring the smooth running of the site at all times. Interns Tylor Birthisel, Melissa Fredericks, Melinda Hurlbut, and Sarah Gibson (formerly Spears), and the site's dedicated crew of volunteers and Utah Friends of Paleontology members, performed the bulk of the work of discovering, excavating, and preparing all the SGDS fossils—without their invaluable work, the SGDS would be much poorer and simply couldn't function today. In fact, the SGDS would not have been possible without these people.

We are extremely grateful to rising-star paleoartist H. Kyoht Luterman for working with us to create the most accurate, up-to-date, and, in some cases, first-ever life restorations of many of the SGDS plants and animals, and to Anna Oakden for her expert photography of the SGDS fossils for this publication.

Undoubtedly we have forgotten to acknowledge someone whose efforts have benefited the SGDS. Rest assured that the oversight was not intentional, and you most certainly have our gratitude.

FOSSIL LAWS

Public Lands

In the United States, most fossils are found on land managed by government agencies for use by all people; such land is called "public land." The managing agencies are either federal, such as the Bureau of Land Management (BLM), the US Forest Service (USFS), and the National Park Service (NPS), or state, such as state park systems and other agencies. In the American West, and especially in Utah, vast tracts of these public lands have exposures of fossil-bearing strata. Anyone enjoying these lands therefore has a great potential to discover fossils weathering out of the rocks. Paleontologists certainly encourage people to look for fossils because there are far too few professional paleontologists available to find all the fossils that are out there. However, to ensure that fossils found on public lands are used responsibly and preserved for *all* people, several laws have been enacted that determine what a person can and cannot do legally with a fossil found on public land. The laws have many things in common but differ slightly from one kind of land to another:

- The collection of fossils of any kind is illegal on any NPS lands (all national parks and monuments) without a valid NPS permit.

- Any land from which fossils are collected must be only minimally disturbed, and cannot be subject to excavation except by permit (and even then, the land usually must be restored after any excavation). The use of power tools or heavy equipment is forbidden, again except by permit, and on some kinds of lands, forbidden altogether.

- In Utah, a rockhounding permit (for which there is a small annual fee) is required to collect legally on most state lands, except state parks, where all collecting is illegal. Other states may have different laws for their own public lands.

- Reasonable, small quantities of fossils of plants and invertebrates may be legally collected on BLM or USFS lands, but collected specimens cannot be sold or bartered.

- The collection of fossil vertebrates (including both body and trace fossils, and the replication of tracks) on any federal public land is illegal without a permit, and permits are issued only to qualified scientists. All fossil vertebrates collected must be preserved in a government-approved repository, usually a museum or university, where they can be properly treated, cared for, and made available for legitimate scientific research and public education. Vertebrates are treated differently than plants and invertebrates because their fossils are typically much rarer. If you do happen to find vertebrate fossils on any public lands, *leave the fossils alone* but report them to the land managers and/or paleontologists and geologists at a local museum, college, or university. If you can, take some pictures of the fossils and the area in which they are found. If you have a GPS unit, the coordinates at which the fossils are found are also very helpful.

Anyone caught collecting fossils illegally or destroying fossils on public lands can be fined or jailed—and yes, such cases have been prosecuted successfully. More information about the specific rules and regulations on any particular type of land can be obtained from your local land management office (BLM, USFS, NPS, etc.). These agencies are generally eager to help you collect fossils enjoyably, safely, and legally. Some helpful websites to consult include:

- "Hobby Collection," Bureau of Land Management, http://www.blm.gov/wo/st/en/prog/more/CRM/paleontology/fossil_collecting.html
- "Laws and Policy," Bureau of Land Management, http://www.blm.gov/wo/st/en/prog/more/CRM/paleontology/paleontological_regulations.html

- "Hobby Collecting of Fossils and Petrified Wood," US Forest Service, http://www.fs.usda.gov/detail/r3/recreation/regulations/?cid=fsbdev3_022262
- "Permits," National Park Service, http://www.nature.nps.gov/geology/permits/index.cfm
- "Rockhounding Utah," Utah.com, http://www.utah.com/hike/rock_hounding.htm
- "Rockhounding Permit Program," School and Institutional Trust Lands Administration, http://trustlands.utah.gov/download/mining/Rockhounding.pdf

Private Lands

Fossils found on private lands are treated very differently than those found on public lands: what happens to those fossils is entirely up to the landowner. This is true whether the fossils are of plants, invertebrates, or vertebrates. The landowner could donate them to a museum, give them away, sell them, or even destroy them. Of course, if the fossils are donated to a university or museum, then they are available for thousands, even millions, of people to enjoy in perpetuity. If the fossils belong to a previously unnamed organism, then it could be named for the generous landowner or collector. If you want to collect fossils legally from private land that is not your own, you must have prior permission from the landowner(s); otherwise, you will be trespassing and stealing.

INTRODUCTION
CHAPTER 1

Southwestern Utah is famous as a beautiful desert of deep red rocks, surrounded by lofty mountains and the towering sandstone cliffs of Zion National Park. Utah, and the rest of North America, sits well north of the equator, in a band of generally temperate climate; the American Southwest, however, is largely desert because the Sierra Nevada and other mountain ranges in California tend to prevent moisture coming from the Pacific Ocean from reaching the continental interior, creating an extensive rain shadow. But like every other landmass, Utah has shifted positions continually over time, the result of plate tectonic activity, and in each position has experienced different environments: it has been a tropical sea, a range of rolling mountains, a vast expanse of sand dunes, and an enormous lake.

At the very end of the Triassic Period, a little over 200 million years ago, the land that would eventually become southwestern Utah was completely different from the mountainous desert it is today. North America, which was still joined to all the other modern continents in the last great supercontinent of Pangaea (fig. 1.1), was much closer to the equator, so much of it enjoyed a wetter, more tropical climate. The mountains of what would eventually become the western United States had not yet been

uplifted, so southwestern Utah was very flat, and the western North American shoreline lay in what is today eastern California, not far from the Nevada border. The area was covered by a dense forest through which large rivers flowed, and seasonal monsoons occasionally produced enormous floods.

Very shortly after the start of the Jurassic Period, roughly 200 million years ago, things began to change. As Pangaea began to break apart, North America drifted northward, out of the band of tropical climates and into more arid zones that were only seasonally wet. Southeastern Utah (and adjacent parts of Arizona, New Mexico, and Colorado) were covered by a vast erg resembling parts of today's Sahara, Namib, and Taklamakan Deserts. Southwestern Utah bordered this erg but was not (yet) covered by this "sea" of sand dunes. Instead, the area contained a series of rivers and floodplains that drained into a large lake centered around today's Pipe Spring National Monument, Arizona, and that extended, at its maximum, northward to near Cedar City, Utah, eastward nearly to central Arizona, and westward into part of southern Nevada as well (fig. 1.2). A wide variety of animals and plants lived in and around this water source, and their story is preserved today as a remarkable series

1

A

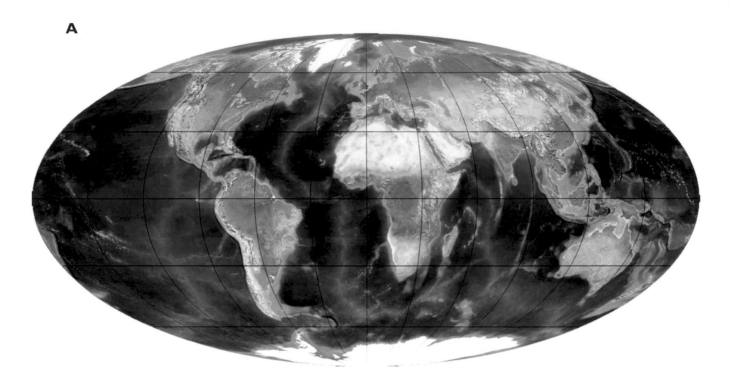

1.1. *A*, Map of the world with the continents arranged as they are today. *B*, Map of the world with the continents arranged as they were at the beginning of the Jurassic Period, approximately 200 million years ago, when the great supercontinent of Pangaea, made up of all of today's separate continents, was just beginning to break apart. *(Maps by Ron Blakey, Colorado Plateau Geosystems, Arizona; used with permission.)*

B

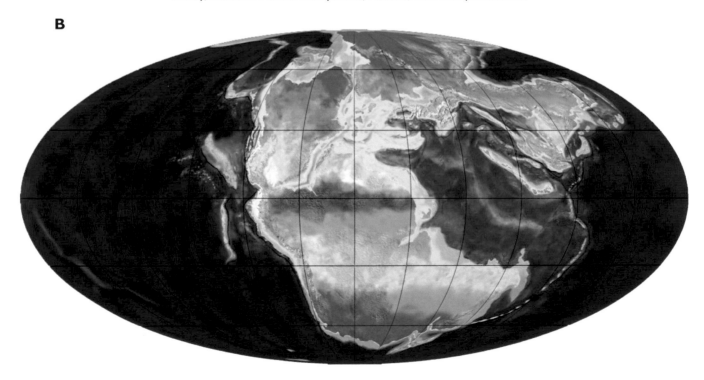

of fossils that were discovered completely by chance. These fossils provide a "snapshot" of life at the beginning of the Jurassic Period with unusual fidelity. Most are footprints and other traces left by living things, but fossils of the bodies of many of the organisms themselves are also present. The footprints made by dinosaurs are the most famous.

The Discovery of the Tracks

On February 26, 2000, Dr. Sheldon Johnson, a retired optometrist and longtime resident of the city of St. George in Washington County, Utah, was leveling a hill on his property on the south side of Riverside Drive for future development (fig. 1.3A). Most of the rock layers that made up this hill were shale and mudstone, which were easy to break up and remove with his trackhoe. However, near the bottom of the hill, Dr. Johnson struck a thick layer of sandstone that did not break apart easily. Fortunately (from his perspective), the layer was dissected by a number of parallel joints, so using cranes and other heavy machinery, Dr. Johnson began the task of lifting out enormous, multiton sandstone blocks. During one such move, one of the blocks accidentally flipped over, exposing a stunning treasure: a perfect natural replica of a dinosaur footprint (fig. 1.3B). It was so well preserved that for a moment, Dr. Johnson thought he had found an actual dinosaur entombed upside down, with its foot sticking out of the rock.

With this revelation, Dr. Johnson began examining other sandstone blocks and discovered that the undersides of almost all of them were covered with dinosaur tracks and other peculiar structures. Purely by accident, he had discovered an incredible treasure trove of fossils, right within St. George city limits (fig. 1.4A). Dr. Johnson wisely contacted local geologists and paleontologists, who quickly recognized the importance of the discovery.

Within just a few weeks, word of the discovery spread and the site received a great deal of public and media attention. By late June, more than fifty thousand people had

1.2. Map of what is today the American Southwest, including Utah and parts of Colorado, New Mexico, and Arizona (dashed lines = state borders), as it was at the beginning of the Jurassic Period, roughly 200 million years ago. Freshwater Lake Dixie, in which sediments were deposited that today make up the Whitmore Point Member of the Moenave Formation, covered what is now southwestern Utah and northwestern Arizona and abutted the vast desert of the Wingate Erg to the north and east. (Map by Ron Blakey, Colorado Plateau Geosystems, Arizona; used with permission.)

visited the site. The discovery also prompted closer investigation of other, nearby properties that bore outcrops of the same layers of rock that Dr. Johnson had uncovered on his property. Inevitably, these investigations led to many more discoveries, not only in the same sandstone layer, but in other layers both above and below the one in which Dr. Johnson's tracks were preserved. Subsequent

1.3. *A,* Dr. Sheldon Johnson in 2000 using a trackhoe to excavate property on what is today the St. George Dinosaur Discovery Site at Johnson Farm (SGDS). *(Still photo from the museum's introductory video presentation.) B,* Dr. Sheldon Johnson in 2000 with one of the first dinosaur tracks he discovered during his excavation. *(Photo by Jim Kirkland, Utah Geological Survey; used with permission.)*

work on Dr. Johnson's property continued to produce new discoveries that attracted still more national and international media attention, including traces made by a crouching, meat-eating dinosaur. Dr. Jim Kirkland, the state paleontologist of Utah, declared the site "the most significant tracksite in western North America."

Creating a Museum

Dr. Johnson and his wife, LaVerna (a retired schoolteacher), worked with the City of St. George to ensure that the site could be preserved and protected for the benefit of future generations. They envisioned a museum being built at the site at which people could learn about the fossils and their meanings. To assist in this effort, geologists and paleontologists from the Utah Geological Survey, the University of Colorado at Denver Dinosaur Trackers Research Group (led by Dr. Martin Lockley, a world authority on fossil footprints), Dixie

State College (now Dixie State University) in St. George, and many other institutions worked closely with the Johnsons and the City of St. George to study the site's treasures. In October 2001, the City of St. George appointed Andrew R. C. Milner as city paleontologist (at the time, one of only three city paleontologists in the world) to preserve, prepare, and document the more than 1,200 tracks that had been discovered by that time. He painstakingly detailed the tracks through tracings, measurements, photography, and casting. In 2005, the Johnsons' dream was realized as a new, permanent museum building (fig. 1.4B) was erected over the original track site, which was officially named the St. George Dinosaur

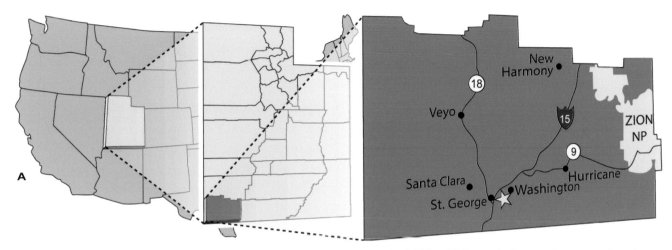

1.4. A, Location of the SGDS in Washington County, Utah. B, Museum at the SGDS, which was built over the exposed track surface to protect and house its fossils. (Photo by Jerry D. Harris.)

Discovery Site at Johnson Farm (SGDS for short).

Since Dr. Johnson discovered that first dinosaur track, paleontological exploration of the area and the rock layers in Washington County has turned up not only more tracks, of both dinosaurs and other animals, but also other, equally important fossils. These include plants, invertebrates, an enormous collection of fossil fishes, and even a few actual dinosaur bones, many in close association with tracks. Tracks and fossils from the bodies of extinct animals are found together only very rarely—usually, the processes necessary to preserve plants, shells, and bones are not conducive to preserving tracks, and vice versa.

Why Is the Discovery Important?

Dinosaur tracks are not a new discovery. They have been found all over the world in rock layers deposited throughout the entire Mesozoic Era (fig. 2.1A), from 253 million to 66 million years ago. The Mesozoic is sometimes referred to as the "Age of Dinosaurs," though dinosaurs were only a tiny fraction of all the different kinds of organisms that lived during this time.

Some of the first dinosaur tracks ever documented were discovered in 1802 in Massachusetts. Many of these early tracks from Massachusetts and elsewhere in New England are preserved today at the Beneski Museum of Natural History at Amherst College. However, they were not recognized as dinosaur tracks until much later because at that time, few dinosaur bones were known and the concept of "dinosaur" had not yet been defined: the first descriptions of dinosaur bones were published in 1824, but the word "dinosaur" wasn't even invented until 1842. Nevertheless, further study over the next several decades made the New England tracks a global standard for dinosaur track studies, particularly for tracks from the period during which they were made. The New England tracks are similar to, and of similar age as, the ones at the SGDS.

Even among the numerous other dinosaur tracks found worldwide since the 1800s, the SGDS tracks are unusual in many ways. Most obviously, many are extraordinarily well preserved, with much greater clarity and detail than most other dinosaur tracks. In addition, an unusually large number of different kinds of tracks and traces are found together within a small area and were made over a short period—essentially, the remains of an entire ancient ecosystem are preserved at the SGDS and around Washington County, Utah. Furthermore, some of the traces were made by tiny, delicate animals such as insects, whose tracks and traces rarely fossilize. Sites with such a diversity of fossils of different organisms are very rare. Finally, the fossils capture a much wider range of track-maker behaviors than normal, providing fascinating insights into the lives of long-extinct organisms.

GEOLOGY

CHAPTER 2

Sedimentary Rock

Most of the rock layers in and around St. George are made of sedimentary rock, meaning rock made of what was, when originally deposited, loose sediment. Sedimentary rock is one of three rock types that geologists recognize. The other kinds are igneous rock and metamorphic rock. Igneous rock solidifies from magma (liquid rock underground) or lava (liquid rock at the surface); some familiar types of igneous rock include granite and basalt. Metamorphic rock forms when other types of rock are buried deeply underground or are caught in zones where continents collide; both situations create either high temperature or high temperature plus high pressure conditions that alter the chemical structures of the original rock to create new types of rock. One familiar type of metamorphic rock is marble, which used to be limestone before being altered by high temperatures.

Some kinds of sedimentary rock, such as limestone, rock salt, and rock gypsum, formed in and around ancient oceans. But most of the sedimentary rock in and around St. George is made of sediment that was deposited on land. These sediments typically came from natural weathering (breaking down) of exposed, preexisting rock in upland areas; gravity, aided by flowing water and wind, invariably pulled the broken-off particles downhill into low-lying areas called basins.

Sediments that are pieces of preexisting rock, and the sedimentary rock made of such sediments, are categorized by grain size. Conglomerate is a sedimentary rock made up of rounded sediment grains of many different sizes, ranging from boulders down to microscopic, mud-sized particles. Sandstone is made of sand—geologically speaking, "sand" means any grains that are 0.0025–0.0790 in. (0.06–2.00 mm) in size. At the small end of that range, the grains are barely visible to the naked eye, but the rock still feels rough to the touch. Sediment made of particles smaller than sand size is called mud—note that mud, geologically speaking, has nothing to do with water content or stickiness. Siltstone is made of the large mud-sized grains, which are 0.00015–0.00250 in. (0.0038–0.0635 mm) and require a magnifying glass to see. Mudstone and shale are made of the finest mud-sized grains, which are smaller than 0.00015 in. (0.0038 mm) and require a microscope to see. These particles are so small that mudstone and shale feel smooth to the touch. Shale breaks apart into flat, paper-thin pieces, whereas mudstone breaks apart into irregular masses. These categories are important because as a general rule, the size of the grains that make up sedimentary rock provides information about what processes carried them from their source to their present location and how far they may have traveled.

The grains that make up all these types of sedimentary rock around St. George started out as part of much larger bodies of preexisting rock that were exposed in distant mountains (which themselves have long since weathered away), primarily the Appalachian Mountains in eastern North America, but also some highlands to the southwest, around what is today the Arizona-Mexico border. Ice, liquid water, wind, and gravity slowly weathered those mountains into smaller pieces, which were carried by ancient rivers or blown by wind into what is now southwestern Utah and deposited bit by bit in vast, sheet-like layers.

Over time, all this loose sediment was buried under newer sediment and lithified, partly by the weight of newer, overlying sediment compacting the older, underlying sediment, and partly by later groundwater moving through the tiny spaces between the grains and depositing minerals that cemented the grains together. All sedimentary rock, regardless of type, is deposited in layers called strata (stratum when there is just one); the study of how these layers formed and how they are ordered is a branch of geology called stratigraphy. Strata are deposited from the bottom upward as older material is buried by newer, so in general, older rock layers lie under younger ones. Exceptions can occur when the action of plate tectonics tilts, folds, and breaks strata over wide areas; one effect is the pushing of older layers on top of or above younger ones along a fault. This is not a big issue right around St. George, however, except along the Hurricane Fault to the east, which has

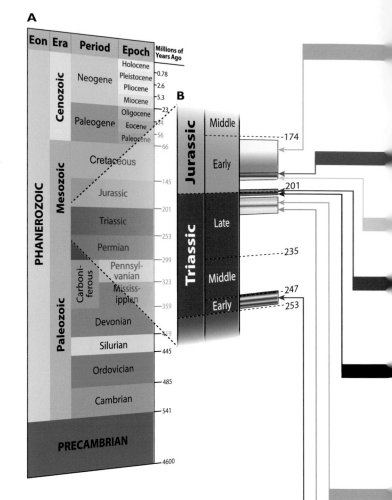

2.1. *A*, Geologic time scale, emphasizing the Phanerozoic Eon (the last one-eighth of Earth's history). *B*, Close-up of the early part of the Mesozoic Era, including the Triassic Period and first half of the Jurassic Period. Most of the rock layers (strata) exposed in the St. George area were deposited within this time frame. *C*, Typical exposures and appearances of the strata in the St. George area, indicating when each layer was deposited. Moenkopi Formation and Shinarump Conglomerate at Quail Creek State Park; Shinarump Conglomerate along Highway 91 just west of Kayenta; Petrified Forest Member in Santa Clara; Dinosaur Canyon Member south of the SGDS; Whitmore Point Member and Springdale Sandstone Member along south Bluff Street in St. George; Kayenta Formation along Interstate 15 in northern St. George; Navajo Sandstone along Highway 18 at the north end of Snow Canyon State Park. (Photos by Jerry D. Harris.)

C

Navajo
Sandstone
Formation

Kayenta
Formation

Springdale
Sandstone Member,
Kayenta Formation

Whitmore Point
Member, Moenave
Formation

Dinosaur Canyon
Member, Moenave
Formation

Petrified Forest
Member, Chinle
Formation

Shinarump
Conglomerate
Member, Chinle
Formation

Shinarump
Conglomerate
Member, Chinle
Formation

Moenkopi
Formation

pushed older oceanic limestone above younger sandstone, siltstone, and mudstone.

Geologic Time

Over the last two centuries, geologists have constructed and refined the geologic time scale, a system for dividing Earth's history into meaningful parts. The basic divisions are two eons, the Phanerozoic and the Precambrian. Each eon is divided into smaller eras; eras are divided into still smaller periods; periods are divided into still smaller epochs, and even epochs can be divided into stages. The basis for divisions of the Phanerozoic Eon, the time during which macroscopic life became abundant on Earth, is evolutionary and extinction events (see "Highlight: Life through Time" at end of this chapter): bigger extinction events that led to more pronounced evolutionary events are the basis for larger divisions, such as between eras and periods; smaller extinction events that led to less pronounced evolutionary events are the basis for smaller divisions, such as between epochs and stages. These events did not happen regularly, so the divisions of the Phanerozoic Eon are irregular in length—that is, not all eras, periods, epochs, or stages are the same length. When the time scale was first established, life was virtually unknown in the Precambrian Eon, so its divisions have been based largely on geological events rather than evolutionary and extinction events, although important evolutionary events did indeed happen during this time. Nevertheless, the Precambrian Eon is nearly seven times as long as the Phanerozoic Eon!

Most of the strata immediately around St. George were deposited in a variety of environments during the Mesozoic Era of the Phanerozoic Eon, which began 253 million years ago and ended 66 million years ago. The

Mesozoic Era is divided into three periods: the Triassic (253–201 million years ago), Jurassic (201–145 million years ago), and Cretaceous (145–66 million years ago). The Triassic and Jurassic are further subdivided into Early, Middle, and Late epochs; the Cretaceous has only Early and Late epochs (fig. 2.1*A*; "Highlight: Life through Time").

In and around St. George, most of the strata range in age from Early Triassic through Early Jurassic, though there are gaps even within that (fig. 2.1*B*). Older strata, from the Paleozoic Era (541–253 million years ago), which preceded the Mesozoic, are exposed south and west of St. George in the Virgin River Gorge and the Beaver Dam Mountains, where they sit on even older, Precambrian igneous and metamorphic rocks. Younger rocks, from the Middle Jurassic and Cretaceous, as well as from the Cenozoic Era (66 million years ago to today), which followed the Mesozoic, are exposed north of St. George around the Pine Valley Mountains, north and east of Zion National Park, and east of Cedar City on the Markagunt Plateau. The youngest rocks in the St. George area are basalt lava flows that erupted from many cinder cone volcanoes beginning about 2 million years ago; the last eruptions were a mere 32,500 years ago (to learn how geologists determine the age of rocks and fossils, see "Highlight: How Do We Know How Old Rocks and Fossils Are?" at the end of this chapter). These very young lava flows sit right on top of rocks that are 5,400–6,000 times older!

St. George Area Stratigraphy

The strata that produce most of the fossils in and around the SGDS are part of the Moenave Formation, which is divided into two parts (figs. 2.1*C*, 2.2). The upper (and therefore

younger) part of this rock unit, in which the majority of the fossils at the SGDS are found, is the Whitmore Point Member (fig. 2.1*C*), which is about 82 ft. (25 m) thick. The lower (and therefore older) part of the Moenave Formation is the Dinosaur Canyon Member (fig. 2.1*C*), which is about 158 ft. (48 m) thick and is made of layer upon layer of mud, silt, and sand deposited at irregular intervals in stream channels and floodplains that covered the area at the very end of the Triassic. The sediment that would eventually become the Whitmore Point Member of the Moenave Formation was deposited during the very beginning of the Jurassic in a vast, long-gone lake geologists today call Lake Dixie (fig. 1.2). The stream- and lake-deposited sediments in the Moenave Formation have since been lithified into shale, mudstone, siltstone, and sandstone.

In southwestern Utah and northwestern Arizona, the boundary between the Triassic and Jurassic Periods lies somewhere within the Moenave Formation, though its exact position is a matter of some debate and current research. Some geologists have placed it within the Dinosaur Canyon Member; others have placed it near the middle of the Whitmore Point Member. Because of this imprecision, it is not clear whether most of the fossils at the SGDS are from the latest Triassic or earliest Jurassic, but in this book we generally consider them earliest Jurassic (roughly 200 million years old).

The Moenave Formation isn't the only set of strata in the St. George area. Below (and therefore older than) the Moenave Formation are strata of definite Triassic age. In the Early Triassic (about 250–245 million years ago), a quiet, very shallow extension of the Panthalassic Ocean covered southern Utah. At this time, North America was close to the equator, so this shallow sea lay under a hot sun that frequently evaporated the seawater, depositing gypsum and limestone along with the sand, silt, and mud carried into the sea by streams. This environment was similar to that of the modern-day Persian Gulf. Today, the colorful red, orange, and gray strata that compose the remains of this sea, and the tidal flats bordering it, are part of the Moenkopi Formation; these strata are particularly well exposed northeast of St. George around Quail Creek Reservoir (fig. 2.1C), in the southern suburbs of Bloomington and Bloomington Hills south of St. George, and in the eastern part of the city of Washington northeast of St. George. Fossils are not common in the Moenkopi Formation around St. George, though some shells of sea animals and a few footprints of land animals have been found.

After the Moenkopi sea retreated, no new sediments were deposited in the St. George area for about 20 million years, spanning most of the Middle Triassic and into the Late Triassic. Then, not long into the Late Triassic (from about 225 to 210 million years ago), a series of very large rivers, surrounded by vast floodplains, began depositing thick packages of sediment, a process that continued for several million years. These sediments are preserved today in the Chinle Formation. The bottom of the Chinle Formation is a thick conglomerate-and-sandstone layer called the Shinarump Conglomerate Member (fig. 2.1C); this layer forms the prominent, tilted ridge that runs along the southern edge of St. George and its western suburb, Santa Clara. Above the Shinarump is the Petrified Forest Member, named for its most famous exposures in Petrified Forest National Park, Arizona. The colorful red, purple, and gray shales and mudstones of the Petrified Forest Member (fig. 2.1C) in Arizona and southeastern Utah

contain a great deal of petrified wood, as well as fossil leaves, shells, and bones of many different kinds of plants and animals that lived in the lush, dense forests surrounding the giant Chinle rivers. Fewer fossils have been found in this member in the St. George area, although it has yielded several important discoveries in recent years. Much of the cities of St. George and Santa Clara are built directly on the soft, easily eroded rocks of this stratum (fig. 2.1). In addition to the sediment that the Chinle rivers deposited in the Petrified Forest Member, erupting volcanoes far to the south and west of the area blew in large quantities of volcanic ash. The remains of that ash decayed over time into certain kinds of clay minerals (locally called "blue clay") that cause no end of headaches for residents whose houses sit on the Petrified Forest Member. These clay minerals expand when wet and contract when dry, slowly flexing and cracking house foundations and causing continual landslides.

The Moenave Formation lies on top of the Chinle Formation. Like the Chinle, most of its soft, easily weathered rocks are buried underneath St. George, but a few outcrops are visible on some hillsides. Sediments of the Dinosaur Canyon Member (fig. 2.1C) began to be deposited on floodplains by rivers at the very end of the Triassic. Fossil plants and tracks have been found in the Dinosaur Canyon Member in St. George; bones of dinosaurs and small, early relatives of today's crocodiles and alligators have been found in this set of strata on the Navajo Nation in northern Arizona (particularly in Dinosaur Canyon, for which it was named). The types of fossils found so far in the Whitmore Point Member indicate that the upper portion of the Moenave Formation was deposited during the first division of the Early Jurassic Epoch, the Hettangian Stage (201.3–199.3 million years

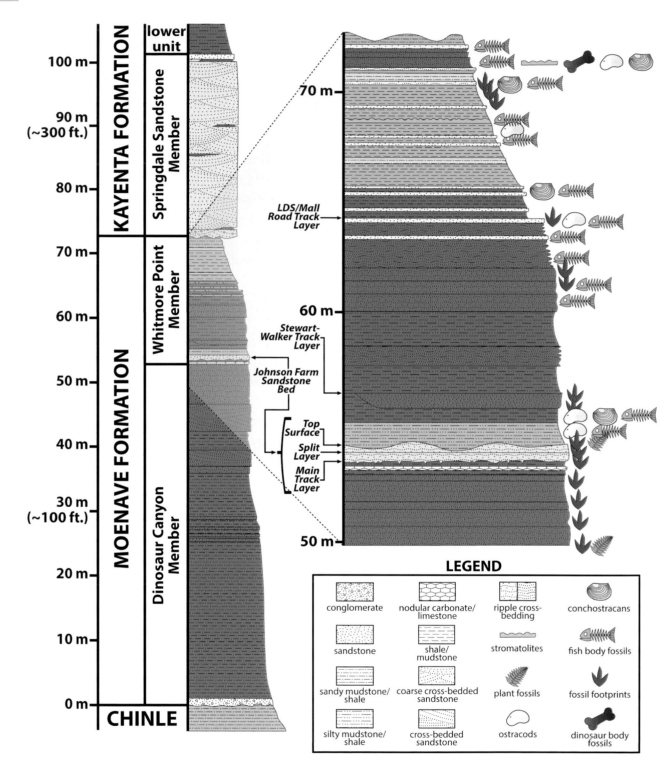

2.2. Details of Moenave Formation strata exposed at and around the SGDS. Multiple layers have produced fossils of various organisms. The natural casts on the overturned blocks on display in the museum (for example, fig. 1.3*B*) are from the Main Track Layer of the Johnson Farm Sandstone Bed, while the natural molds on display on the undisturbed surface within the museum (for example, fig. 5.1*B, C*) are on the Top Surface of the Johnson Farm Sandstone Bed.

ago). Lake Dixie lasted for at least a few million years during this stage, alternately growing and shrinking over time. It may have lasted even into the next part of the Early Jurassic, the Sinemurian Stage (199.3–190.8 million years ago). At the same time that Lake Dixie covered southwestern Utah, a huge erg deposited sand in massive dunes in southeastern Utah (fig. 1.2). Those dunes are preserved today as the Wingate Sandstone Formation, which forms many of the towering sandstone cliffs in the Four Corners area of the Southwest.

After Lake Dixie dried up for the final time, stream, floodplain, and pond deposits returned to the area; these are preserved in deep red rocks of the Early Jurassic (about 198–195 million years old) Kayenta Formation, which sits on top of (and is therefore younger than) the Moenave Formation. The beautiful red ridges surrounding St. George are exposures of the Kayenta Formation (fig. 2.1C). The Kayenta Formation also preserves lots of dinosaur footprints, in both the thick Springdale Sandstone Member at its bottom and the siltier layers above it. Some of these can be seen in Warner Valley, southeast of St. George, and around the city of Washington, northeast of St. George.

Eventually, the Kayenta streams were overtaken by vast, new ergs that are preserved as the Early Jurassic Navajo Sandstone Formation, an extremely thick set of strata. Around St. George, it makes up the red and beige cliffs of Snow Canyon State Park (fig. 2.1C) and the bluffs of the Red Cliffs Desert Reserve, but it is even more famous for forming the bulk of the spectacular towering cliffs of Zion National Park. Fossils of dinosaurs and other animals are rare in the Navajo Sandstone Formation—good examples of fossil footprints can be seen near the Subway, a popular hiking spot in Zion

National Park—but the formation is largely not fossiliferous. The age of the Navajo Sandstone Formation is uncertain because fossils and other datable material (see "Highlight: How Do We Know How Old Rocks and Fossils Are?") are so rare in it, but its deposition probably began about 195 million years ago and lasted until near the end of the Early Jurassic Epoch.

Younger, Middle Jurassic and Cretaceous strata overlie the Navajo Sandstone Formation; exposures of these rocks ring the Pine Valley Mountains and can be seen north of St. George, especially around the communities of Diamond Valley and Gunlock. Lava flows, the youngest rock in the area, can be seen all over Washington County; some of the surviving cinder cones that produced the flows can be seen in Diamond Valley and around the city of Hurricane, northeast of St. George.

Stratigraphy of the Moenave Formation at the SGDS

New research on and in the Moenave Formation around St. George, inspired by Dr. Johnson's discoveries, has produced many new finds. The Moenave Formation in the St. George area was previously largely ignored by paleontologists, but it has now proven to be quite rich in fossils. In and around St. George, fossil tracks have been identified in twenty-six different layers of the Moenave Formation, from the uppermost part of the Dinosaur Canyon Member and all through the Whitmore Point Member (fig. 2.2). Many of these fossils have been mapped as they occur in the ground because it is impossible to remove the enormous slabs in which they occur.

The most important track-bearing layer is a layer of sandstone 12–28 in. (30–70 cm) thick that contains the tracks first discovered by Dr. Johnson, which was named the Johnson

Farm Sandstone Bed in his honor. Although it is difficult to see in most places, the Johnson Farm Sandstone Bed is itself actually made up of multiple layers of sandstone, each deposited at a different time. We know this because some of the blocks at the SGDS have split apart along these other layers, revealing still more footprints *within* the Johnson Farm Sandstone Bed. Footprints could have been made only on sandy surfaces exposed to the air, so the sediments of the Johnson Farm Sandstone Bed could not have been deposited all at once; therefore, even it can be subdivided. The layer at the bottom that contains the now-famous tracks is called the Main Track Layer (fig. 2.2). One of the more productive layers (in terms of tracks) above the Main Track Layer, but still within the Johnson Farm Sandstone Bed, is called the Split Layer (fig. 2.2) because it must be carefully split open to see the tracks it preserves. Still more footprints are found at the top of the Johnson Farm Sandstone Bed in what is called the Top Surface Layer. Like the Split Layer, the Top Surface Layer itself comprises many thin strata; tracks occur on several of these, but most are on a single surface. The SGDS museum was built over this surface (blocks that were not turned over by Dr. Johnson), and it forms the floor of about half of the building.

Other tracks have been found in still higher layers of the Whitmore Point Member. One locality, the Stewart–Walker Tracksite, was named in honor of Darcy Stewart, a prominent St. George business owner who has donated significant specimens and time to the SGDS and who owned the land on which these tracks were found, and former SGDS coordinator Theresa Walker, who discovered the site. Approximately sixty tracks occur at this site on four different stratigraphic levels, each separated by about 19.7 in. (50 cm) of rock (fig. 2.2). These beds are about 4.9 ft. (1.5 m) above, and therefore slightly younger than, the Top Surface Layer of the Johnson Farm Sandstone Bed. Rare tracks of plant-eating dinosaurs were discovered in one of these levels.

Still more tracksites lie in strata above the Stewart–Walker Tracksite. Two of these are the LDS and Mall Drive Tracksites, named for the former landowners (the LDS Church) and a nearby road, respectively. At present, over 3,000 tracks, each in its exact position and stratigraphic context, have been mapped at all of these Whitmore Point Member tracksites. Well over 4,000 more tracks are preserved on the blocks from the Main Track Layer and other Moenave Formation strata around Washington County. Undoubtedly, thousands more tracks and other fossils have yet to be discovered.

Highlight: Life through Time

The fossil record of life, though incomplete because of the spotty nature of the fossilization process, provides a remarkably vivid portrait of life through time. This diagram shows the positions of some of the major evolutionary events in Earth's history on the geologic time scale. Note that even though the Precambrian Eon (from 4.6 billion to 541 million years ago) appears short on the time scale, seven-eighths of Earth's history is included in that time span.

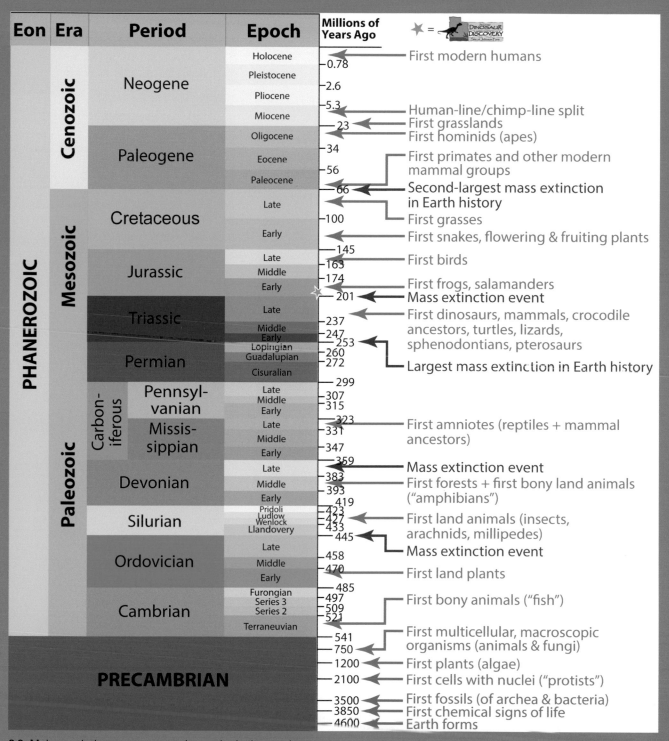

2.3. Major evolutionary events on the geologic time scale.

Highlight: How Do We Know How Old Rocks and Fossils Are?

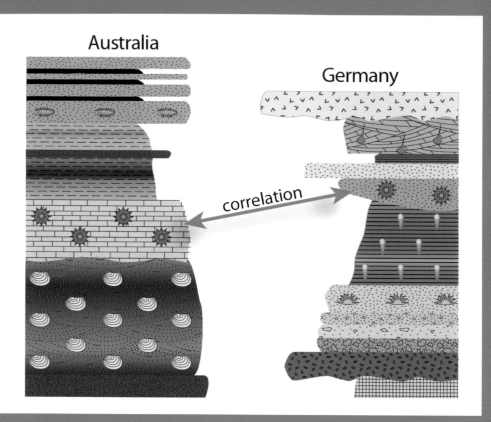

2.4 A. How stratigraphic dating works.

Geologists have several different ways of figuring out how old a rock or fossil is. The two most common methods are called stratigraphic dating (formerly called relative dating) and radiometric dating (formerly called "absolute" dating).

Stratigraphic dating is perhaps the simplest method, but it usually relies on fossils being present in the stratum being dated. If fossils are present, a geologist can compare them to fossils found in rocks elsewhere in the world. If the fossils are the same as those found elsewhere, then those rocks, and the fossils in them, must be roughly the same age in both places. This is because evolution happened in the past, just as it happens today. Because organisms evolve and change through time, most species (at least, of animals) live for only a short time—typically just a few million years, and often less. Therefore, fossils of identical organisms, even if found in far-flung places, must be roughly the same age (give or take up to a few hundreds of thousands of years). This system works so well that geologists are able to define zones of specific sets of fossils that are unique to certain time periods. Rocks that contain identical fossils are said to correlate with each other (fig. 2.4A). Paleontologists depend on such correlations to understand how organisms changed through time in different places around the world.

Stratigraphic dating is a powerful tool, but it doesn't tell a geologist the age of a rock or fossil in actual number of years. To obtain an actual number, geologists turn to radiometric dating. This method takes advantage of the fact that many—but not all—rocks contain radioactive atoms of different elements. A radioactive atom is one that is unstable because it contains an imbalance of protons and neutrons in its nucleus.

(The number of protons is important because that's what defines each element.) The unstable atom spontaneously tries to become stable by either ejecting particles from the nucleus or absorbing other particles into the nucleus—these are the processes of radioactive decay, which in some instances produce radiation that can be harmful in large doses. After decaying, the atom has a new number of protons in its nucleus, so it belongs to a daughter element, which is different from its original, unstable, parent element. For example, ^{235}U (uranium-235, so called because it has 235 particles in its nucleus: 92 protons and 143 neutrons)

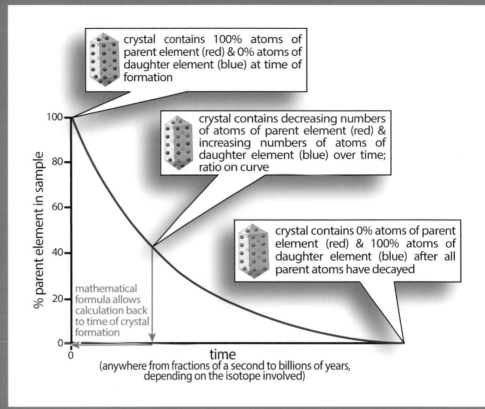

decays over several steps into ^{207}Pb (lead-207, so called because it has 207 particles in its nucleus: 82 protons and 125 neutrons). Although the decay is spontaneous for any one atom, it happens at a regular rate for each isotope; the rates for different isotopes range from fractions of seconds to billions of years. For rocks containing radioactive atoms of an element, geologists can use special machines to count how many parent and daughter element atoms are present in any single crystal in the rock. All the daughter atoms used to be parent atoms before they decayed, so the age of the rock is the time when there were all parent, and no daughter, atoms. The ratio of the number of parent and daughter atoms can then be combined with the known decay rate for the parent element, and the age of the rock can be easily calculated (fig. 2.4*B*).

2.4 *B*. How radiometric dating works.

With a few exceptions, radiometric dating can be performed only on igneous rock, specifically on individual crystals of certain minerals inside an igneous rock. Igneous rock forms when magma (melted, liquid rock) solidifies into interlocked crystals of various minerals. In liquid form, all the atoms in the magma float freely and are not bound to one another. When the magma solidifies, however, some atoms will bond to certain other atoms, forming crystals of different minerals. In most cases, parent and daughter elements do not bond in the same minerals, so when a datable crystal forms, it contains only parent atoms (fig. 2.4B). But some of those atoms will, over time, decay and transform into daughter atoms—no daughter atoms can enter from the outside because they are all locked up in other crystals of other minerals. With rare exceptions, radiometric dating cannot be used on fossils because fossils are the remains of living organisms, which cannot survive in magma where the datable crystals form. Radiometric dating can, however, be used on rocks containing fossils *if* the rock contains any igneous material, such as a volcanic ash, that was deposited along with sediment. Fossils in ash-bearing sedimentary rocks are therefore highly prized.

But some rocks have neither fossils of widespread organisms nor bits of igneous rock in them and therefore cannot be dated either stratigraphically or radiometrically. However, if any overlying or underlying rocks have either stratigraphically or radiometrically datable material, then they can provide relatively narrow age ranges for the strata in between. A stratum of rock *must* be younger than any stratum below it, and it *must* be older than any stratum above it—remember, strata are deposited from the bottom upward and are therefore oldest at the bottom and youngest at the top. Having determined the ages either stratigraphically or radiometrically of a huge number of strata around the world, geologists have many points of comparison that they can use to "sandwich" any stratum that's free of igneous material between overlying and underlying rocks with known ages, ideally within a narrow range.

As an example, Figure 2.4C shows how this process can be used to date a dinosaur bone in a set of strata in the western United States. Here, neither the purple dinosaur bone nor the stratum it is in can be dated directly, and the age range (167–83 million years old) provided by the overlying and underlying datable strata is too broad to be very useful. However, the ages of some of the overlying and underlying fossils can be determined based on the ages of other strata in other parts of the world. In Brazil, a stratum contains the same fossil as one that underlies the dinosaur bone in the western United

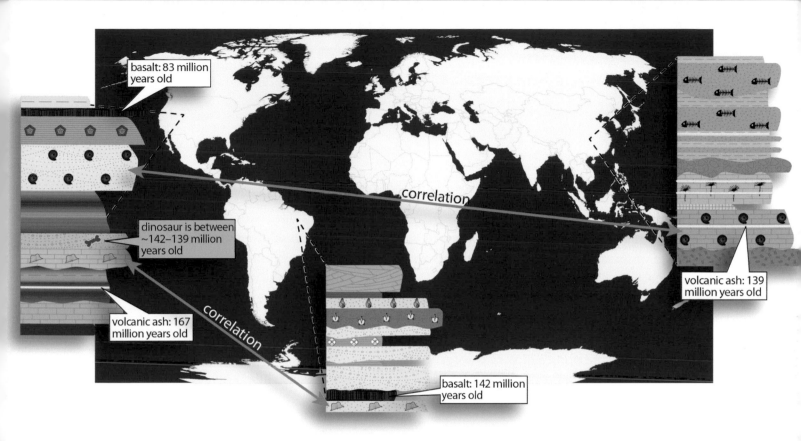

basalt: 83 million years old

correlation

dinosaur is between ~142–139 million years old

volcanic ash: 139 million years old

volcanic ash: 167 million years old

correlation

basalt: 142 million years old

States. In Brazil, this fossil underlies a datable basalt; because it underlies that basalt, it must be the same age or older than the basalt (it must be 142 million years old or older). That minimum age of 142 million years also applies to the fossil in the western United States. Similarly, in China, a stratum contains the same fossil as one that overlies the dinosaur bone in the western United States. In China, this stratum contains a volcanic ash that can be dated. Because that ash is in the same rock as the fossil, that age applies to the fossil, too (they were deposited at the same time), and to the fossil in the western United States. The purple dinosaur bone is now "sandwiched" between two well-dated, other fossils, and its age thus falls within a much narrower range (142–139 million years old) than it did based only on the data from its surrounding strata. This is why geologists from around the world publish their fossil finds and dates, creating a useful library of this information for future geologists There aren't many rock layers left anywhere in the world for which a narrow age range is still unknown.

2.4 *C.* How stratigraphic and radiometric dating techniques work together to help determine the age of a fossil, even when the fossil has no known correlate anywhere in the world and is not in a rock that can be dated radiometrically.

TRACE FOSSILS

CHAPTER 3

Types of Fossils

All fossils (see "Highlight: What Is a Fossil?" at the end of the chapter) can be placed into one of two categories: body fossils or trace fossils. Most of the fossils with which people are familiar, such as those of bones, teeth, shells, leaves, and petrified wood, are body fossils—they were once actual parts of the bodies of various organisms. Trace fossils, on the other hand, are fossils that were *made* by an organism's body but were never *themselves* parts of the organism's body. In short, they are traces of the organism's passing or behavior. (Note that *any* kind of organism can leave a trace—most familiar are traces of animals, but plants, fungi, and even single-celled organisms such as bacteria can leave traces, too.) The most common, and perhaps easiest to understand, type of trace fossil is a track: tracks were made by a part of an animal's body (a foot or hand), but the tracks themselves were never parts of the track maker's body. Body parts other than hands and feet (bellies, tails, etc.) can also leave traces (usually called imprints or impressions, rather than tracks). Other trace fossils include animal burrows that have been filled in with sediment, root casts (spaces made by plant roots that have become filled with sediment), coprolites (fossil feces),

nests, and bite marks on bones and shells made by the teeth of other organisms. Fossil eggs straddle the boundary between trace fossils and body fossils: the egg is not a part of the body of the mother that laid it, but in some ways the eggshell can be considered a part of the body of the embryo inside the egg until it hatches, so some paleontologists consider fossil eggs to be body fossils, while others consider them trace fossils.

Most of the fossils found at and around the SGDS are tracks and, therefore, trace fossils. Many of the tracks are in trackways (a row of tracks made by the same individual track maker). While tracks may not seem as impressive as fossil bones—people love to go to natural history museums in large part to see fossil skeletons—tracks are very important to understanding ancient animals. One reason is that an organism generally has only one body, and that body has a very small chance of ever becoming a fossil; even plants, which shed their leaves and grow new ones throughout their lives, have a limited supply of potential fossils. But an organism that can move around, especially an animal, can leave tens of thousands of tracks and traces, each one of which has the potential to become a fossil. Trace fossils therefore are generally far more common than body fossils. A second, more important reason is that tracks were made by *living* organisms, but skeletons and other body parts, when found as fossils, are the remains of *dead* organisms. To understand what an extinct organism was like when it was alive and to learn how it behaved on a day-to-day basis, paleontologists study trace fossils because they capture behaviors as they happened.

Like body fossils, tracks have to be buried quickly in order to be preserved as fossils. Sediment washed into a track helps protect the track from being destroyed. If a track was made in loose sediment (such as dry sand, silt, or mud), the action of new sediment washing or blowing over and into it normally would partly or completely destroy the track, and it would not be preserved. However, if a track was made in stickier sediment (such as wet sand, silt, or mud) that later dried out, the sediment grains would stick together and sometimes become crusty. Some kinds of bacteria also help bind sediment grains together by precipitating certain minerals from the water in the spaces between sediment grains. Such crusty sediment can become more resistant to weathering by wind or currents of water, and the tracks become miniature basins that may fill with sediment before they can be destroyed. Such conditions are not very common, which is why not all footprints ever made are preserved. But these must have been the conditions in which the SGDS tracks, especially the very detailed ones, formed.

Natural Molds versus Natural Casts

Tracks can be preserved in two different ways: as natural molds and natural casts (fig. 3.1). Natural molds are the simplest to understand: they are the footprints themselves, the indentations made by an animal's appendage, such as your foot might make walking in loose sediment on a beach or along a riverside. When an animal steps into sediment, the sediment molds itself around the appendage to create a kind of "negative," or impression of the appendage. The finer the sediment, the more detailed the track. If the sediment in which a natural mold is made is undisturbed before it is buried by more sediment, it has the potential to be preserved as a fossil.

Natural casts are created when new sediment fills in the natural mold: the sediment occupies the same space as the appendage that made the track. When lithified, the natural cast becomes an exact natural replica (a cast) of the track-making animal's appendage made entirely of sediment. Paleontologists often make artificial casts of tracks, too, by pouring rubber, plaster, or plastic resin into natural molds, a complex procedure that requires special techniques to avoid damaging the natural mold. Too often, unsupervised amateurs try to make replicas themselves and damage, or even destroy, tracks. On public land, such attempts are actually illegal—see "A Note to Readers: Fossil Laws" for more information.

The dinosaur tracks that Dr. Johnson found on the undersides of his sandstone blocks were natural casts; unfortunately, the process of removing the blocks on which the casts occurred tended to destroy the natural molds they filled on a more fragile mudstone layer underneath. When the Johnson Farm Sandstone Bed is carefully split to expose the Split Layer, both natural molds and the natural casts that filled them can be seen (the molds on one side of the split and the casts on the other; fig. 3.1). The tracks on the Top Surface, at the top of the

3.1. How natural molds (m), undertracks (u), and natural casts (c) of dinosaur (and other fossil animal) tracks formed and are preserved. *A,* A dinosaur walks in loose sediment during the Mesozoic Era, creating tracks (natural molds). The weight of the dinosaur also compacts and deforms several buried layers of sediment (visible in the cross sections), creating undertracks (close-up). *B,* The tracks sit exposed until the next time sediment is deposited in them. If there is no deposition in a short period—hours, days, or weeks—the tracks will likely be either weathered away before they can be preserved, or destroyed when other organisms disturb the track-bearing sediment. *C,* During a river flood, lake expansion, or other process, new sediment is deposited inside the track. This sediment forms a replica, or natural cast, of the dinosaur foot that originally made the track. *D,* Over millions of years, all the layers of sediment, including the ones that hold undertracks, natural molds, and natural casts, compact still further and lithify (become solid rock) when they are buried deeply under the weight of later, overlying layers of sediment. The layer of sediment containing the natural casts of the dinosaur footprints can be split from the layer containing the natural molds (close-up). *(Restoration by H. Kyoht Luterman.)*

Johnson Farm Sandstone Bed, are all natural molds; at the SGDS, the natural casts that once filled them in were inadvertently destroyed in the leveling of the hill that once covered them, but some can still be seen in the same layer elsewhere in St. George.

When a track is made, not only the sediment right at the surface is affected: the weight of an animal can drive its appendage deep into the sediment, deforming even older, already-buried layers, although the deformation is not as strong as in the actual track. Thus, even layers under the one the track-making animal stepped in can preserve a version of track called an undertrack (fig. 3.1). Undertracks get lithified and preserved along with the actual tracks; when the strata bearing the actual tracks weather away, undertracks can still remain. Because the sediment containing the undertracks was less disrupted than the sediment that contained the actual tracks, an undertrack can look rather different from an actual track—for example, an undertrack can have narrower, shallower digit impressions that have different proportions than those of the actual track. Sometimes, undertracks are merely shapeless, bowl-like structures. Undertracks can therefore appear to have been made by a different animal than the one that made the actual track. Even trained paleontologists sometimes have difficulty determining whether a track is an actual natural mold or an undertrack.

Life in Moenave Time

Many different kinds of organisms lived in and around Lake Dixie. Animals walking in the soft, loose sediment in and around its shores created the traces now preserved as fossils. When animals living in the lake died, their bodies sank to the bottom; sometimes, parts of land plants and animals also got washed into the lake. Occasionally, these remains were buried quickly (before they could decay completely) and eventually became body fossils. In lakes, the remains of dead organisms can be buried quickly and often, so ancient lake deposits, such as those in the Whitmore Point Member of the Moenave Formation, are especially prized by paleontologists because they are often better-than-usual places to find fossils.

Dinosaur tracks and other fossil footprints are well known in Early Jurassic–age rocks of the southwestern United States, especially in the Kayenta Formation, but skeletal remains are rare in the region. Until the discovery of the SGDS, only a relatively small number of tracksites had been documented in southwestern Utah, or in the Moenave Formation as a whole. Not only did the SGDS discoveries fill a gap in the fossil record of this area, but the local interest they generated resulted in the discovery and reporting of numerous additional sites in the region, especially by SGDS staff and volunteers. These sites include multiple track and bone localities in the Late Triassic–age Chinle Formation (under the Moenave Formation) and in the Early Jurassic–age Kayenta Formation (atop the Moenave Formation) (fig. 2.1).

Ichnology

The study of tracks and traces, called ichnology, is a specialization within paleontology. Like all sciences, ichnology has its own language of technical terms. A small but important part of ichnology is naming and classifying trace fossils (technically called ichnites). Naming footprints and other traces may seem a strange practice, but scientists do it for the same reason that they name organisms themselves: to make communication between scientists easier. Each name describes a trace with a particular set of characteristics. However, in order to understand trace fossils better, it is critical to recognize that the names given to trace fossils are *entirely separate* from the names of the organisms that made them. That is, the name of an organism is *not* the same as the name of a trace it might make, and vice versa. The name of a trace generally says nothing about which organism made it. The reason for this unusual system is that many different organisms are capable of creating traces that are indistinguishable from one another, especially when fossilized. For example, imagine trying to distinguish the footprints of a sparrow from those of a finch without having seen which bird made the tracks. These are clearly different animals, but they have virtually identical feet and therefore create identical tracks. The same is true for most extinct organisms: many different kinds were capable of creating identical traces. We cannot go back in time and observe an organism making a track, so it is impossible to be certain that one particular kind of organism, especially one that is extinct, made any specific track. Worse, because body fossils are so rare, it is possible—even probable—that the maker of a particular trace fossil hasn't yet been discovered from body fossils.

To get around these problems, different kinds of fossil tracks are given their own scientific names using a system similar to the one used for living organisms. Every organism is a member of a species. For example, humans belong to the species *Homo sapiens*, domestic dogs belong to the species *Canis familiaris*, and the common American crow belongs to the species *Corvus brachyrhynchos*. (The terms "human," "dog," and "crow" are common names; many species, particularly extinct ones, such as *Tyrannosaurus*

rex and *Homo erectus*, have only scientific names and no common names.) Two or more species are grouped together into a genus—*Homo, Canis, Corvus,* and *Tyrannosaurus* are genera (the plural of genus). Genera group together several closely related and very similar species. For example, *Canis latrans* (coyote), *Canis lupus* (wolf), and *Canis aureus* (jackal) are species that are different from (but closely related to) *Canis familiaris*. Trace fossils are treated the same way, except instead of being put into species and genera, they are put into ichnospecies and ichnogenera (ichnogenus for just one). Each ichnospecies and each ichnogenus has certain distinctive characteristics that ichnologists use to name and classify them.

The practice of naming and characterizing distinct types of tracks and traces was initiated by Reverend Edward Hitchcock in the 1830s. Hitchcock, the father of modern ichnology, studied Late Triassic and Early Jurassic tracks from New England from the 1830s through the 1850s. Most of the tracks, and certainly the most famous ones, that Hitchcock studied were made by vertebrates, though he studied traces left by invertebrates, too. Hitchcock's life's work was published in 1858; it was revised by paleontologist Richard Swann Lull in 1904 and 1953, and dozens of new vertebrate track types have been named by other paleontologists since then.

Below, we will use the ichnological approach to introduce the SGDS trace fossils. First we will examine the different types of traces found at the SGDS, but without getting into what organism may have made each kind of trace; we will return to discuss the trace makers afterward. In the meantime, look at the pictures of each type of trace and think about the descriptions. Can you hypothesize from the pictures and descriptions what kind of organism made each kind of trace?

Cast of Characters: Who's Who at the SGDS
Vertebrate Tracks and Traces

Many of the tracks that Reverend Hitchcock studied and named in the 1800s from New England are similar or identical to those found at and around the SGDS, so the names Hitchcock created apply at the SGDS, too. The most common vertebrate tracks in New England, as well as at the SGDS, are tridactyl and were made by bipedal animals. Hitchcock noticed that these tracks were somewhat similar to tracks made by many kinds of birds: the middle of the three toes was the longest. The tridactyl tracks Hitchcock observed came mostly in two common varieties: small tracks with relatively skinny toe impressions and shorter outer toes, and large tracks with more robust toe impressions and outer toes that were closer in length to the middle toe. If all these tracks had been identical except in size, Hitchcock might have given them all one name, but he discerned more subtle differences between them, involving features such as toe thickness, toe length, the angles between toes, and how the toe pads were arranged on the foot. Based on these kinds of differences, he gave them different names accordingly.

Hitchcock named the smaller (typically 6 in. [15 cm] or less in length), skinnier tracks *Grallator*. He named the larger (typically 10 in. [25 cm] or more in length), thicker-toed tracks *Eubrontes*. *Grallator* and *Eubrontes* are ichnogenera, and many ichnospecies have been named in each ichnogenus based on even more subtle differences. However, there never were (nor will there ever be) animals named *Grallator* or *Eubrontes*—remember, these are the names of *track* types, not the names of the animals that made them. In 1904, Lull gave the name *Anchisauripus* to tridactyl tracks that were between *Grallator* and *Eubrontes* in size. But it isn't clear whether these are really just larger *Grallator* or smaller *Eubrontes* tracks, so many paleontologists no longer use the name *Anchisauripus*. Some paleontologists have hypothesized that *Grallator* tracks were made by relatively small animals of one species, *Anchisauripus* by medium-sized animals of a different species, and *Eubrontes* by large animals of a third species. However, it is also possible that two, or even all three, track types were made by the same kind or kinds of animals as they grew from small, young individuals to large adults. Both scenarios are probably correct in different cases.

Grallator is the most common vertebrate track type at the SGDS (figs. 3.2A, B). *Eubrontes* is also present (fig. 3.2C), but less common; tracks that some might call *Anchisauripus* are rare. Tracks similar to *Eubrontes* but that are tetradactyl, called *Gigandipus*, are also present but very rare (fig. 3.2D). A few tracks found near the SGDS that are similar in size to *Eubrontes*, but that have skinnier toes and two outer toes that are even closer in length to the middle toe, are called *Kayentapus* (fig. 3.2E). *Kayentapus* tracks are rare in the Moenave Formation, but, as the name suggests, very common in the overlying Kayenta Formation, in which *Eubrontes* and *Gigandipus* tracks are less common. *Grallator*, *Eubrontes*, *Gigandipus*, and *Kayentapus* were all made by obligate bipeds.

A rare track type found at the SGDS belongs to the ichnogenus *Anomoepus* (figs. 3.3A, B). *Anomoepus* is another ichnogenus initially named by Rev. Edward Hitchcock for tracks from New England. Large *Anomoepus* tracks

3.2. Theropod dinosaur tracks at the SGDS. *A*, Natural cast of a typical *Grallator* track (SGDS 197B) made by a small, bipedal, carnivorous dinosaur. This track overlies another, older *Grallator* track: its heel can be seen in the lower left, and the tip of its middle toe is peeking out in the lower right. *B*, Natural mold of a rare, very small *Grallator* track (uncataloged, from the Top Surface) made by either a very young individual of the same kind of dinosaur that made the larger track in *A*, or a very small species that had an identical foot shape. *C*, Natural cast of a well-preserved, detailed *Eubrontes* track (SGDS 9) made by a large, bipedal, carnivorous dinosaur. *D*, Natural cast of a rare *Gigandipus* track (SGDS 50). Note the thicker toes and small but prominent first toe (arrow). *E*, Natural cast of a typical *Kayentapus* track (uncataloged) made by a large, bipedal, carnivorous dinosaur. Note that *Gigandipus* tracks include a trace of the small first toe, whereas tracks of *Eubrontes*, *Grallator*, and *Kayentapus* do not. Pennies (*A–B*) and quarters (*C–E*) for scale. *(Photos by Jerry D. Harris.)*

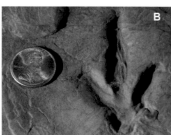

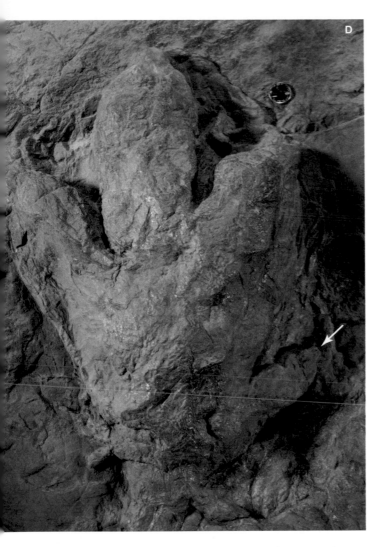

are similar in size to tridactyl *Grallator* tracks, but in *Anomoepus*, the toes are arranged differently and are of different proportions than in *Grallator*. Some *Anomoepus* tracks were made by bipedal track makers, who left only footprints, while others were made by quadrupedal track makers, who left both footprints and handprints. Some *Anomoepus* trackways include both foot-only and hand-and-foot prints, indicating that at least some *Anomoepus* track makers were facultative bipeds. *Anomoepus* footprints are tridactyl, but the handprints are tetradactyl or pentadactyl. *Anomoepus* tracks at the SGDS are nearly always

tridactyl and bipedal (fig. 3.3*A*), though rare handprints have also been discovered (fig. 3.3*B*). *Anomoepus* tracks are known from several places around the world, but the SGDS specimens are some of the oldest. Few *Anomoepus* tracks have been found in the Johnson Farm Sandstone Bed; most of the specimens come from slightly higher layers in the Whitmore Point Member of the Moenave Formation, especially the Stewart–Walker Tracksite surfaces, and from the Kayenta Formation elsewhere in Washington County (fig. 3.3*A*).

Fairly common at the SGDS are small tracks made by obligatorily quadrupedal, tetradactyl vertebrates. Hitchcock named similar tracks from his New England sites *Batrachopus* (figs. 3.3*C, D*). Rarer tracks at the SGDS are similar to *Batrachopus* but are normally pentadactyl and have longer fingers and toes in a somewhat different arrangement. These tracks may belong to a poorly known ichnogenus called *Exocampe* (fig. 3.3*E*), though the SGDS tracks do not always preserve five digits. The rarest track type at the SGDS, of which only two or three specimens have been found so far, also resembles *Batrachopus* and *Exocampe* but is broader and larger and has shorter toes in a different arrangement (fig. 3.3*F*); this track type has not yet been named, but it resembles an ichnogenus called *Brasilichnium*.

The size range of vertebrate tracks from the SGDS is remarkable. The smallest *Grallator* tracks are only about 0.8–1.2 in. (2–3 cm) long; the largest, which are rare, are around 10 in. (25 cm) long, larger than usual for *Grallator* tracks, but these big specimens retain the proportions of smaller *Grallator* tracks rather

than those of the more robust *Eubrontes* tracks. *Eubrontes* tracks exceed 12.6 in. (32 cm) in length, and *Kayentapus* tracks are roughly as big. Only a very few tridactyl tracks fall in the intermediate, "*Anchisauripus*" range of 9.8–12.6 in. (25–32 cm). This separation of tridactyl track sizes suggests that *Grallator* and *Eubrontes* tracks were made by at least two, and maybe more, different species of track makers. If these tracks were all made by different-aged individuals of the same species, the absence of tracks in one size range would mean that

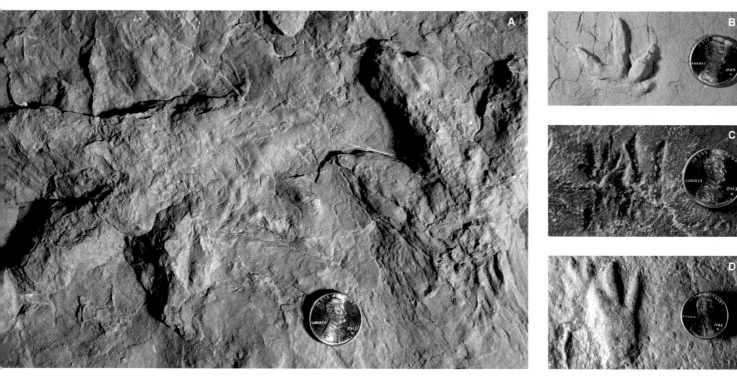

3.3. Tracks made by vertebrates other than theropod dinosaurs at the SGDS. *A*, Three-toed *Anomoepus* footprints (SGDS 1296) made by small, bipedal, herbivorous ornithischian dinosaurs. Unlike most of the tracks at the SGDS, these are from the Kayenta Formation, not the Moenave Formation. *B*, Natural cast of a rare tetradactyl *Anomoepus* handprint (SGDS 166). *C–D*, Sets of *Batrachopus* handprints and footprints, made by a small, quadrupedal, early relative of modern crocodiles and alligators. *C*, Natural molds of right hand and wrist (smaller, outside) and foot (larger, inside) prints (SGDS 170), dusted with chalk for highlighting. *D*, Natural casts of left hand (smaller, outside) and foot (larger, inside) prints (SGDS 570). *E*, Two pairs of rare, possible *Exocampe* tracks (arrows; SGDS 509) that may have been made by a small, quadrupedal, lizard-like sphenodontian. Note that in each pair, both handprints (smaller, to the outside) and footprints (larger, to the inside) are preserved. Next to the reptile tracks is a typical *Parundichna* trace (paired wavy lines between arrowheads; SGDS 509) made by the paired fins of a swimming fish, possibly a small coelacanth. *F*, A rare, as yet unnamed type of track (SGDS 215), probably made by a medium-sized "protomammal." *G*, A typical *Undichna* trace (SGDS 917), made by the tail of a swimming fish. Pennies (*A–E, G*) and quarter (*F*) for scale. (*Photos A–E by Jerry D. Harris; F–G by Anna Oakden.*)

individuals of that age did not frequent the area even though younger and older individuals both did—possible, but not very likely. Nevertheless, the larger *Eubrontes* track makers had to have started life as small babies, so undoubtedly some of the *Grallator* tracks were also made by young *Eubrontes* track makers, too.

Batrachopus tracks typically fall in the size range of 0.4–2.0 in. (1–5 cm) in length, though larger tracks, up to about 3.1 in. (8 cm), are known. The hind footprints of *Batrachopus* are larger than the handprints; the hind footprints

are also more commonly preserved than the handprints because the hind limbs bore more of the track maker's weight and therefore sank into the sediment more deeply than the hands and forelimbs. Also, the hind footprints may partly or completely cover the handprints on occasion, a phenomenon known as overprinting, meaning that the track maker's hind foot stepped on the same spot where the hand made a print just moments before, destroying part or all of the handprint. One SGDS *Batrachopus* track maker traveled 17 ft. (5.2 m) along the top of a ridge on the Top Surface (fig. 2.2).

Though tracks made by larger, land-dwelling vertebrates, such as *Batrachopus*, *Grallator*, and *Eubrontes*, are impressive and attractive, they are not the only vertebrate traces at the SGDS. The underwater environment of Lake Dixie had an abundance of smaller, perhaps less noticeable (but certainly just as important), vertebrates that also left interesting trace fossils. Traces consisting of single thin, wavy lines, sometimes broken into crescents, go by the name *Undichna* (fig. 3.3*G*); paired thin, wavy lines go by the name *Parundichna* (fig. 3.3*E*). These traces are very unlike toe-bearing tracks because they were made by very different kinds of vertebrates.

3.4. Fossilized burrows at the SGDS. *A, Scoyenia* burrows (on block SGDS 797), possibly made by beetle larvae. *B,* Filled-in, vertical *Skolithos* burrows (SGDS 504) made by either worms or spiders. The burrows protrude into the rock; here you see only their tops. *C, Helminthoidichnites* burrows (SGDS 741), possibly made by small worms. *D, Palaeophycus* burrows (SGDS 191), probably made by large worms. Pennies for scale. *(Photos by Anna Oakden.)*

Invertebrate Tracks and Traces

A vast number of small invertebrates that probably formed the base of the food web in and around Lake Dixie in the Early Jurassic also left trace fossils at and around the SGDS. Burrows are the most common and abundant type of trace fossils, both at the SGDS and in sedimentary rocks around the world. Short, linear, horizontal burrows of the ichnogenus *Scoyenia* (fig. 3.4*A*) occur at the SGDS, often in abundance, in the upper 10 ft. (3 m) of the Dinosaur Canyon Member of the Moenave Formation and in at least four layers in the lower part of the Whitmore Point Member. Abundant long, vertical, cylindrical *Skolithos* burrows (fig. 3.4*B*) cover enormous areas in association with other tracks and fish fossils in Whitmore Point Member strata above the Johnson Farm Sandstone Bed. Two of the best

D

Skolithos-bearing layers are called the Slauf Burrow Bed and Sally's Burrow Bed. These two beds are named after their discoverers, David Slauf and Sally Stephenson, both of whom are dedicated volunteers at the SGDS and members of the Southwestern Chapter of the Utah Friends of Paleontology. The Slauf Burrow Bed also preserves some dinosaur tracks and fish fossils.

Long, skinny, randomly winding burrows of the ichnogenus *Helminthoidichnites* (fig. 3.4C) differ from short, straight *Scoyenia* burrows. A fourth type of burrow, called *Palaeophycus* (fig. 3.4D) comprises thick but simple, branching,

3.5. Trace fossils made by invertebrates at the SGDS. *A, Kouphichnium* trackway (SGDS 258) made by a small horseshoe crab. *B,* Part of a *Diplichnites* trackway (SGDS 1201B), probably made by an arthropod. *C, Bifurculapes* trackway (SGDS 197B), probably made by an adult beetle or a notostracan. *D,* V-shaped traces made either by dragonfly larvae or perhaps by small rocks pushed through the sediment by water currents. If actual trace fossils, these marks would belong in the ichnogenus *Protovirgularia. E,* Probable *Lockeia* traces (SGDS 234), which are resting traces made by small clams, or possibly conchostracans, sitting in the sediment. Pennies for scale. *(Photos A–B, D–E by Anna Oakden; C by Jerry D. Harris.)*

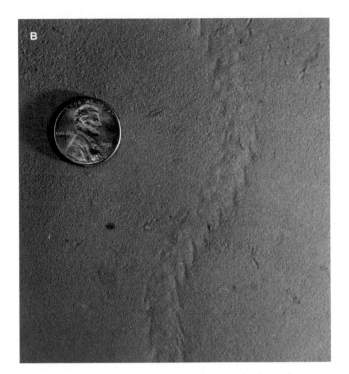

and horizontal burrows. *Palaeophycus* burrows occur at the base of a green sandstone bed in the middle of the Whitmore Point Member.

Tracks made by walking invertebrates are less common than burrows; often they are delicate and not impressed deeply into the sediment and therefore are easily destroyed before they can be buried and preserved, or they are not noticed unless the lighting on a slab of rock is just right to highlight them. Conditions at the SGDS 200 million years ago, however, were just right to preserve many of these traces. Tracks called *Kouphichnium* (fig. 3.5A) resemble multiple parallel rows of lines, some of which have upside-down V shapes on them. *Diplichnites* (fig. 3.5B) trackways, resembling paired rows of dots or dashes, are fairly common at the SGDS. Much rarer *Bifurculapes* trackways look like rows of crescents nested tightly into one another (fig. 3.5C). Some possible traces at the SGDS that resemble sets of Vs nested into one another (fig. 3.5D) may belong to the ichnogenus *Protovirgularia*.

Invertebrate traces are often good indicators of ancient environments, much more so than vertebrate traces are because many invertebrates live only in specific environments. So in addition to filling out what we know of the animals that lived in this area at the beginning of the Jurassic, these traces help tell us which environments (on land near the edge of Lake Dixie or underwater in the lake; in wet versus dry soils and sediments, etc.) were in which places at various points in time. Some are very specific: *Scoyenia* fossils, for example, occur only in what used to be wet soils.

Who Made the Tracks?

As mentioned earlier, it is actually impossible to say with certainty that any particular trace fossil was made by a particular species of organism unless a body fossil of the organism is found directly in the trace fossil. While very rare, such fossils have actually been found with some invertebrates. For example, fossil horseshoe crabs in Late Jurassic–age strata in Germany are preserved sitting where they died at the ends of long, winding trackways. Burrowing animals are particularly susceptible to preservation inside their own traces: flash floods can fill

burrows with sediment before their occupants can escape, so the organisms get buried alive and can drown or suffocate while still in their burrows, where they have a good chance of becoming fossils. Fossil crayfish, lungfish, amphibians, early relatives of mammals as well as true mammals, extinct crocodylians, and even a few kinds of dinosaurs have been found preserved in burrows, though not at the SGDS. But thus far, no dinosaur body fossil has ever been found at the end of a trackway, so associating particular footprint ichnogenera or ichnospecies with particular genera or species of dinosaurian track makers remains impossible.

Nevertheless, narrowing down the maker of a particular trace fossil to a short list of candidates is possible. One method, which is particularly useful for invertebrate traces, is to study the traces made by living organisms and compare them to fossil examples. Most of the groups of invertebrates that are alive today have been around for hundreds of millions of years (although there has been lots of evolution within each of these groups), which permits relatively strong comparisons between today's invertebrates and ancient, fossil examples. However, such comparisons obviously cannot work well for animals that have no living representatives, or whose living representatives have evolved very different body shapes from those of their ancient relatives.

A second method, which is particularly useful for vertebrate traces, is to compare the shape of a track to the shapes of the appendages of various animals. This method is especially useful when comparing the traces to organisms of the same age—for example, if the foot bones from a set of body fossils match the shape of a set of trace fossils in rocks of the same age, then the animal to which the foot bones belong *may* have (but not definitely!) made the trace fossil. But the body fossils and trace fossils must be of about the same age: comparing the forefeet and hind feet of a dog, for example, to a track found in Early Jurassic–age rock wouldn't make much sense because dogs didn't evolve until well into the Cenozoic Era. Similarly, comparing the hands and feet of dinosaurs such as *Tyrannosaurus* and *Triceratops* to tracks at the SGDS doesn't make much sense either because *Tyrannosaurus* and *Triceratops* lived at the end of the Cretaceous, not the beginning of the Jurassic. Restricting such comparisons to animals found in Early Jurassic-age rocks makes much more sense.

Vertebrate Trace Makers

Wavy *Undichna* and *Parundichna* fossils are the kinds of traces made by the fins and tails of fishes while swimming. Although fishes are not typically thought of in association with "tracks," swimming fishes often do leave traces. Some fishes rest or live near the bottoms of rivers, lakes, and oceans, and when they swim, the tips of their fins sometimes drag through the sediment. Fishes swim largely by undulating (waving) their bodies and tails from side to side, so the traces made when their tails drag in sediment look like single, undulating lines—the ichnogenus *Undichna*. Other fishes have long, paired fins on the undersides of their bodies; when these fins drag through the sediment, they make paired, crisscrossing, undulating lines— the ichnogenus *Parundichna*. Many fish body fossils have been found in Lake Dixie sediments (see chapter 4), so these kinds of trace fossils are not unexpected. Many different kinds of fishes were equally capable of creating identical *Undichna* and *Parundichna* traces.

Grallator, *Eubrontes*, and *Kayentapus* tracks are all tridactyl and share certain features of toe shape, length, and arrangement. Feet that match the configuration of these footprints belong to a group of dinosaurs called theropods (fig. 3.6A). Most theropods were carnivores, though a few were herbivores. Some theropods are among the most famous dinosaurs: *Tyrannosaurus*, *Velociraptor*, *Carnotaurus*, and *Spinosaurus* are all theropods. The Late Jurassic theropod *Allosaurus* is even the state fossil of Utah. One group of theropods— birds—is the only group of dinosaurs that survived the extinction event at the end of the Cretaceous, which is often described as having "wiped out the dinosaurs." This understanding is relatively recent, but nevertheless, the idea that dinosaurs went extinct at the end of the Cretaceous is outdated: birds are dinosaurs that live among us today.

Most theropods have tetradactyl feet: they have the same first, second, third, and fourth toes as humans, but the fifth toe was lost through the course of evolution of theropod ancestors early in the Triassic. However, the first toe—the equivalent of the human big toe— in most theropods was very small and did not usually touch the ground. Instead, it sort of hung off the side of the ankle, rather like the dewclaw of a dog or cat (fig. 3.6A). So when theropods walked and ran, they would have made tridactyl footprints, with impressions of the second, third, and fourth toes. These feet bear the features and arrangements seen in the tridactyl tracks at the SGDS. Tetradactyl *Gigandipus* tracks are an exception: these tracks typically include traces made by that small first toe. Perhaps this means that they were made by a kind of theropod that had a particularly long first toe, one long enough to touch the ground when the animal walked. Or, perhaps they are simply tracks made by the same kinds of

3.6. Early Jurassic theropod and ornithischian dinosaur foot shapes. *A*, Skeleton (left) and restorations in top view (middle) and bottom view (right) of the left foot of a typical early theropod dinosaur. Note that theropod feet have four toes, but only three were used in walking and running—the small first toe did not normally touch the ground. *B*, Skeleton (left) and restorations in top view (middle) and bottom view (right) of the left foot of a typical early ornithischian dinosaur. Theropod and ornithischian feet differ in several subtle ways. In the ornithischian, the first toe is longer, the second to fourth toes are more equal in length, the pattern of pads on the bottom of the foot is different, and the claws are blunter than in the theropod. The somewhat longer first toe of the ornithischian (3.3*A*) sometimes touched the ground. *(Restorations by H. Kyoht Luterman.)*

theropods that made *Eubrontes* tracks but with feet that sank more deeply into the sediment, allowing the small first toe to make contact and leave a trace, as well as creating other differences in shape from typical *Eubrontes* tracks. *Kayentapus* tracks appear to have been made by theropods with skinnier toes, and somewhat longer second and fourth toes, than the theropods that made *Eubrontes* tracks.

We can say with a great deal of certainty that the *Grallator, Eubrontes, Gigandipus,* and *Kayentapus* tracks at the SGDS were made by theropod dinosaurs. However, only a few kinds of theropods had evolved by the Early Jurassic. Many of the famous theropods, such as *Tyrannosaurus, Velociraptor, Carnotaurus, Spinosaurus,* and *Allosaurus,* had not yet evolved by this time: *Allosaurus* lived at the end of the Jurassic, almost 55 million years after the time of Lake Dixie, and *Tyrannosaurus, Velociraptor, Carnotaurus,* and *Spinosaurus* all lived at different times in the Cretaceous (fig. 2.1;

B

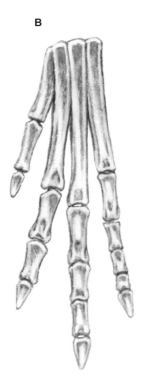

"Highlight: Life through Time"), even longer after Lake Dixie had vanished. In fact, the time span between Lake Dixie and *Tyrannosaurus* (about 134 million years) is more than twice as long as the span between *Tyrannosaurus* and today (about 66 million years). Birds did not evolve until sometime in the Late Jurassic, and no birds big enough to have made tracks in the size range of *Eubrontes*, *Gigandipus*, and *Kayentapus* existed until late in the Cretaceous.

At the beginning of the Jurassic, theropods had been around for only about 30 million years, so the theropods that made the tracks at the SGDS were very primitive compared to their more famous, later cousins. Many of these early theropods belonged to a group called the coelophysoids, most of which were small, long, skinny, quick, and agile predators. The most famous and best-known coelophysoid is *Coelophysis*. Hundreds of skeletons of this theropod have been found in the Chinle Formation in New Mexico and in earliest Jurassic-age rocks in southern Africa. *Coelophysis* may have lived at the time of Lake Dixie, but its fossils have not yet been found in the Moenave Formation. (Some geologists have argued that the part of the Chinle Formation that produces *Coelophysis* fossils is actually the same age as the Moenave Formation, but this has not been universally accepted.) However, a close relative of *Coelophysis*, called *Megapnosaurus* (fig. 3.7A), lived in the same area and at the time Lake Dixie existed. The best-known *Megapnosaurus* fossils come from the Kayenta Formation (which, remember, overlies the Moenave Formation and is therefore younger) in Arizona, but a few bones that seem to belong to this dinosaur have been found in the Dinosaur Canyon Member of the Moenave Formation, also in Arizona. A second, as yet unnamed, small coelophysoid has been discovered in the earliest Jurassic-age Nugget Sandstone Formation (which formed as part of the Wingate Erg) in eastern Utah. Typical small coelophysoids such as these grew to approximately 3.5 ft. (1.1 m) high at the hip and close to 10 ft. (3 m) long. A similar sized but slightly more advanced, non-coelophysoid theropod called *Kayentavenator* also comes from the Kayenta Formation of Arizona.

Grallator tracks may have been made by small coelophysoids, such as *Coelophysis*, *Megapnosaurus*, or the new, unnamed theropod, or even non-coelophysoids, such as *Kayentavenator*. Even though no way exists to determine which theropod or theropods—more

than one kind may have made identical tracks!—made the *Grallator* tracks at the SGDS, *Megapnosaurus* and its relatives are good models for the *kinds* of dinosaurs that would have left at least some of the *Grallator* tracks. The actual track maker could also have been a theropod that hasn't been discovered yet, different from any of the aforementioned ones; this may be particularly true for the larger-than-usual *Grallator* tracks found at the SGDS. And, as mentioned earlier, at least some *Grallator* tracks were probably made by small, young individuals of the animals that would grow up to be *Eubrontes* and other large theropod track makers.

Eubrontes, *Gigandipus*, and *Kayentapus* tracks are too big to have been made by any of the smaller coelophysoids, even the biggest known adults. *Eubrontes* and *Gigandipus* tracks have thick, heavy toes, but *Kayentapus* tracks have skinnier toes of different proportions. At least some *Kayentapus* tracks at the SGDS may be

3.7. *A*, The small coelophysoid theropod dinosaur *Megapnosaurus*. *Megapnosaurus*, and other theropods similar to it, made *Grallator* footprints. *B*, The large theropod dinosaur *Dilophosaurus*. *Dilophosaurus*, and other theropods similar to it, made *Eubrontes*, *Gigandipus*, and *Kayentapus* tracks. (*Restorations by H. Kyoht Luterman.*)

undertracks of *Eubrontes* or *Gigandipus* tracks, but others appear to be actual tracks. This means that at least two different kinds of large theropods, with slightly different feet, roamed the area around Lake Dixie in the beginning of the Jurassic. To date, only one large theropod is known from the Early Jurassic of the American Southwest: *Dilophosaurus* (fig. 3.7*B*), which, like *Megapnosaurus*, comes from the Kayenta Formation of Arizona. However, we know about *Dilophosaurus* feet only from bones, without all the flesh around them, so we don't know whether it had thick, heavy toes that would have made *Eubrontes* and *Gigandipus* tracks, or skinny toes that would have made *Kayentapus* tracks. More importantly, no evidence has yet been found that *Dilophosaurus* itself lived at the time of Lake Dixie. In some studies, *Dilophosaurus* is not a coelophysoid, but instead somewhat more closely related to some of its later relatives, such as *Allosaurus*. Whoever its relatives are, *Dilophosaurus* is a good model for

the kind of dinosaur that was capable of making some of the large theropod tracks at the SGDS. *Dilophosaurus*, as well as any other theropods that made the larger theropod tracks, were 5–6 ft. (1.5–1.8 m) tall at the hip and roughly 20 ft. (6.1 m) long. These theropods were the largest predators of the Early Jurassic, but they would have been only medium sized compared to their truly gigantic, later relatives, such as *Allosaurus* and *Tyrannosaurus*.

Besides the numbers and shapes of toes, toe lengths, and other features, the theropod tracks at the SGDS preserve additional information about the theropods that made them. Some of the tracks were made in sediment of just the right grain size and consistency to capture and preserve actual skin impressions from the feet of the living dinosaurs that made the tracks. This skin was made up of a series of tiny bumps, like the texture of a basketball (fig. 3.8). This was almost certainly *not* the same texture as the skin on the rest of the body: the skin on

B

 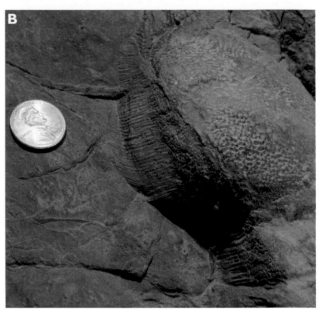

3.8. *A*, Partial (broken) *Eubrontes* natural cast (SGDS 15). Arrowhead indicates region shown in *B*. *B*, Close-up of the side of the *Eubrontes* natural cast in *A*, showing long, skinny scale scratch lines and the impression of the bumpy skin on the sole of the foot. Pennies for scale. *(Photos by Anna Oakden.)*

3.9. The early crocodile relative *Protosuchus*. *Protosuchus* and its relatives made *Batrachopus* tracks. *(Restoration by H. Kyoht Luterman.)*

theropod feet was probably thicker and of a different texture than that anywhere else on the body to aid in gripping the surface, as it is in lizards, birds, and crocodylians. In a few tracks, the skin bumps slid through the sediment while the foot was being impressed into it, carving thin, parallel grooves called scale scratch lines (fig. 3.8*B*). Scratch lines can help determine exactly how the theropod's foot entered, moved through, and exited the sediment; this, in turn, helps paleontologists understand how the track makers walked, which is more difficult to ascertain from skeletons.

Like the theropod tracks, *Anomoepus* tracks are also tridactyl (or occasionally tetradactyl), but they differ markedly in shape and toe arrangement. These tracks better match the feet of early ornithischian dinosaurs (fig. 3.6*B*), all of which were herbivores. The ornithischian dinosaur group includes the armored dinosaurs, such as *Stegosaurus* and *Ankylosaurus*, the horned dinosaurs, such as *Triceratops* and *Protoceratops*, the "bone-headed" dinosaurs, such as *Pachycephalosaurus*, and the "duck-billed" dinosaurs and their relatives, such as *Iguanodon* and *Parasaurolophus*. In the Early Jurassic, however, ornithischian dinosaurs were only just beginning to evolve—there were no horned, "bone-headed," or "duck-billed" dinosaurs yet, and only the most primitive armored dinosaurs. Very few ornithischians lived in the Early Jurassic compared to the later Jurassic and the Cretaceous, and their fossils are more poorly known than those of their later relatives. Most Early Jurassic ornithischians were relatively small, quick, agile, bipedal animals. One of these, a small, early, armored dinosaur called *Scutellosaurus* (fig. 3.10), was found in the Kayenta Formation of Arizona. As with *Megapnosaurus* and *Dilophosaurus*, no evidence has been found that it lived at the time that Moenave Formation sediments were being deposited, but it is a good model for the kind

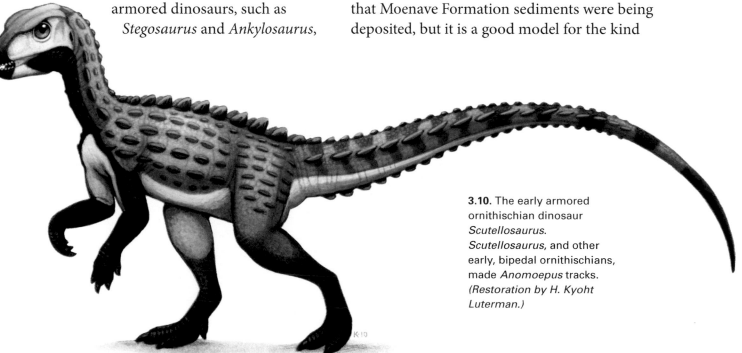

3.10. The early armored ornithischian dinosaur *Scutellosaurus. Scutellosaurus,* and other early, bipedal ornithischians, made *Anomoepus* tracks. *(Restoration by H. Kyoht Luterman.)*

of dinosaur that would have made *Anomoepus* tracks.

Batrachopus tracks were not made by dinosaurs. Rather, *Batrachopus* handprints and footprints better match the skeletons of small, very early crocodylians that were just beginning to evolve in the Early Jurassic. Skeletons of one of these, a small, skinny animal called *Protosuchus* (fig. 3.9), have been found in the Dinosaur Canyon Member of the Moenave Formation in Arizona. Known *Protosuchus* fossils are from individuals a bit too big to have made most of the *Batrachopus* tracks at the SGDS, but the SGDS tracks may have been made by young animals, or maybe by a different, smaller early crocodylian than *Protosuchus*, one that is currently unknown from body fossils. *Protosuchus* and other early crocodylians only vaguely resemble the squat, heavily armored, sprawl-legged crocodylians living today. They had only two rows of thin armor down their backs, and relatively long, upright legs. Instead

of spending most of their time in the water like their living relatives, these early crocodylians were well adapted to running fairly quickly on land. However, some *Batrachopus* tracks at the SGDS show that the track maker sometimes waded into Lake Dixie, possibly to catch invertebrates or fishes for food.

Exocampe tracks, if the SGDS specimens really belong in this ichnogenus, resemble the hand and foot skeletons of members of a group of lizard-like animals called sphenodontians. Only one sphenodontian is alive today: the endangered tuatara of New Zealand. But in the Late Triassic and Early Jurassic, sphenodontians, such as *Clevosaurus* (fig. 3.11A), were very common and lived the lifestyles that many lizards live today. Some Triassic and Jurassic sphenodontians were entirely herbivorous; others ate insects and other small invertebrates.

The rare, unnamed SGDS track type has some similarities to the foot skeletons of what

3.11. *A*, The Early Jurassic sphenodontian (tuatara) *Clevosaurus. Clevosaurus,* and other sphenodontians, may have made *Exocampe* footprints. *B*, The "protomammal" *Kayentatherium. Kayentatherium,* and other "protomammals," made tracks similar to an unnamed SGDS track type. *(Restorations by H. Kyoht Luterman.)*

A

can be called "protomammals" (fig. 3.11*B*)—the kinds of animals from which mammals evolved. Mammals evolved from "protomammals" in the Late Triassic, but some "protomammals" survived and lived alongside true mammals in the Jurassic; some survived even into the Cretaceous in a few places. No "protomammals" are known from the Moenave Formation, but some, including one called *Kayentatherium*, are known from the Kayenta Formation of Arizona. *Kayentatherium* was about the size of a medium-sized dog—quite large for a late-surviving "protomammal"—and is a good model for the kind of animal that made the unnamed SGDS tracks. Other, smaller "protomammals" are known from the Chinle Formation, which underlies the Moenave Formation; if any of these survived into Moenave Formation time, then they may have been responsible for making the SGDS tracks. Quite likely, the track maker was something new and as yet undiscovered.

Invertebrate Track Makers

Helminthoidichnites burrows were probably made by worms, possibly different kinds, because they resemble many modern worm burrows in shape and structure. Most paleontologists think that *Skolithos* burrows were also made by worms (different from those that made *Helminthoidichnites* burrows), but some kinds of spiders also dig burrows that look like *Skolithos* fossils. *Palaeophycus* burrows also may have been made by worms, but different kinds again than those that made *Helminthoidichnites* or *Skolithos* burrows because the traces are fatter and differently shaped. Burrows that also resemble *Palaeophycus* can be produced by some millipedes, but millipedes live in wet, forested areas, unlike the Lake Dixie environment, so the SGDS fossils were probably not made by millipedes, even though millipedes existed in the Early Jurassic. *Scoyenia* burrows may have been made by the larvae of some kind

B

K·12

of insect(s) because they look very much like burrows made by some living insect larvae, particularly some kinds of beetle grubs, but they could also have been made by worms or tiny relatives of crabs and lobsters.

Bifurculapes trackways were also made by insects; in particular, they resemble tracks made by some kinds of living adult beetles. Some kinds of insects, as well as peculiar, pond- and lake-dwelling crustaceans called notostracans, can make traces that closely resemble *Diplichnites*, so the identity of its trace maker is less clear. The V-shaped *Protovirgularia* marks are rather curious: trackways shaped like nested Vs are made by scaphopods, but these peculiar animals live only in salty ocean water and therefore could not have lived in freshwater Lake Dixie. Similar traces are made on land by dragonfly larvae. However, the V-shaped trails at the SGDS also resemble chevron marks, which are trails made by small rocks or other heavy objects when they are pushed through sediment by water currents, so it is not clear whether the *Protovirgularia*-like marks at the SGDS are even trace fossils at all.

The strange, linear rows of tracks called *Kouphichnium* match the tracks made by living horseshoe crabs as they drag their bodies across sediment. Today, horseshoe crabs live only in the ocean, but in the Triassic and Jurassic, many also lived in freshwater. However, the SGDS specimens bear some differences from other known *Kouphichnium* traces. This may be because they were made in a different kind of sediment or under different conditions than other *Kouphichnium* fossils, or they may have been made by an unusual horseshoe crab, or even some other kind of animal—perhaps lobster-like crayfishes (crawdads), whose fossils have been found in Late Triassic–age rocks of the Chinle Formation in Arizona, New Mexico, and Utah.

Some puzzling marks on the Main Track Layer and Top Surface Layer (fig. 2.2) look like small diamonds or wedges (fig. 3.5E). Initially, these were thought to be marks made in the sediment by crystals of certain kinds of salts that precipitated out of Lake Dixie's water when it evaporated under the warm, tropical sun; this phenomenon is common in shallow water bodies in warm climates. If this is correct, it would mean that Lake Dixie was a bit salty—not nearly as salty as the ocean (and certainly not as salty as the Great Salt Lake in northern Utah, which is completely unrelated to Lake Dixie), but even today's bodies of freshwater have some dissolved mineral salts in them. More recently, however, these wedge-shaped marks have been shown to be more similar to marks made by small clams or superficially clam-like animals resting in the sediment, an ichnogenus called *Lockeia*. No actual clam shell fossils have been found in the Moenave Formation, but freshwater clam fossils are known from the underlying Chinle Formation and overlying Kayenta Formation, so it is quite likely that there were clams at the time of the Moenave Formation, too. However, whether clams made the SGDS *Lockeia* traces is uncertain; crustaceans called conchostracans may also have been the culprits (see chapter 4).

What Traces Tell

Simply by virtue of being there, tracks and other traces tell us something about the kinds of environments that the trace-making animals inhabited. After all, animals couldn't make tracks if they didn't at least occasionally venture into the track-making environments. The abundance of the tracks of theropod dinosaurs and early crocodylians at the SGDS means that these animals spent good portions of their time in and around Lake Dixie. Some theropod tracks are known even in what were once the offshore (underwater) parts of the lake, demonstrating that the track makers had no fear of going into the water. Perhaps they frequented the lake to catch and eat the abundant fishes; certainly the fishes were a good food source.

In contrast, the comparative rarity of herbivore tracks, such as *Anomoepus*, in the same strata as the abundant theropod tracks suggests that the herbivorous dinosaurs only rarely came near the water, perhaps only for the occasional drink or bath. They seem to have preferred more distant, upland environments, perhaps where plants (and cover from the predatory theropods) were more abundant. The same may hold true for the sphenodontians and "protomammals," whose tracks are also rare at the site. "Protomammal" tracks and burrows are relatively common in the sand dune deposits that became the Navajo Sandstone Formation, long after the Moenave Formation was deposited, so these kinds of animals apparently did not prefer marginal lake environments.

Walking and Running

Trace fossils are terrific sources of information about how extinct animals behaved because, unlike body fossils, they were made by living animals in the process of behaving. One of the most obvious things trace fossils can tell us about is locomotion—how the track maker moved. This kind of information is very difficult to extract from an unmoving skeleton, so a great deal of what we know about the way many extinct animals moved is based on their footprints and other trace fossils. Even hypotheses about how dinosaurs moved based

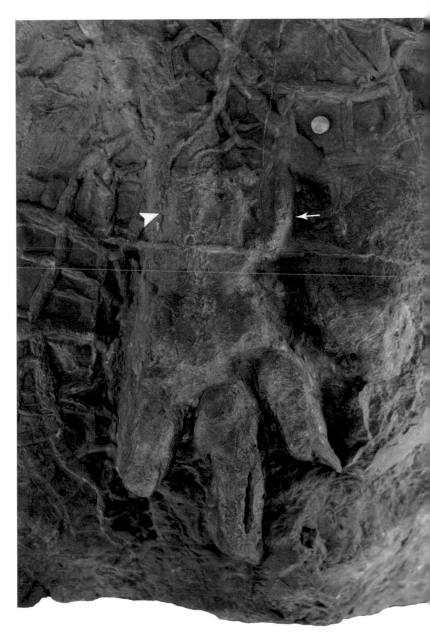

3.12. *Eubrontes* track (SGDS 8) that includes a rare impression of the metatarsus ("heel"; arrowhead), as well as an impression of the small first toe (arrow). This indicates that for this step, the *Eubrontes* track maker—a large theropod—was moving in an unusual way. It may have been crouching to sit or to remain stable while walking on slippery sediment, or it may have slipped a bit in wet sediment and lowered its body to help catch its balance. Penny for scale. *(Photo by Anna Oakden.)*

3.13. Portion of a 26.5 ton (24 metric ton) block (SGDS 568) on display at the SGDS that preserves 58 *Grallator* footprints in 13 trackways. Tracks on this block were made by theropods that were both walking and running. These tracks were not made all at the same time but within a few hours to a few days of each other. The different track sizes show that they were made by several different individuals, possibly of more than one species of small theropod. One foot (30 cm) ruler for scale. *(Photo by Jerry D. Harris.)*

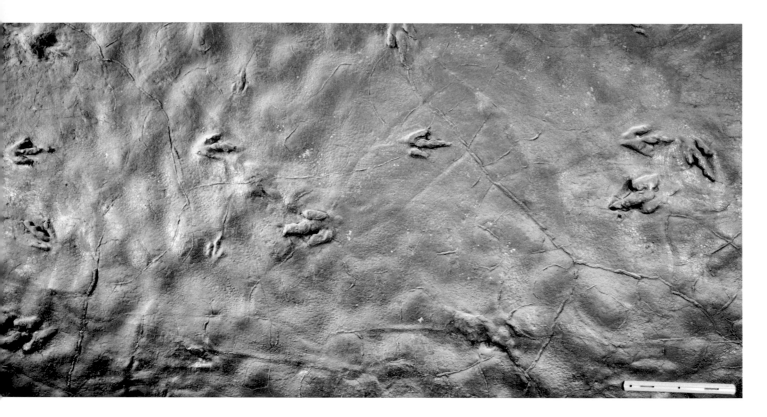

on their skeletons must match up with the fossil track record.

Theropods (including birds) and ornithischians, like all other dinosaurs, were digitigrade, meaning they walked only on their toes, with their ankles well up in the air. Most mammals (including dogs and cats) are also digitigrade, but some, such as deer, cows, and horses, take this to the extreme and are unguligrade, meaning they walk only on the very tips of their toes. In contrast, humans (and only a few other mammals, such as kangaroos and bears) are plantigrade, meaning they walk on the metatarsus, the part of the leg between the ankle and bases of the toes. (For a human, walking in a digitigrade posture

means tiptoeing.) Plantigrade posture brings the ankle into contact with the ground, so plantigrade animal footprints have big heel impressions—we humans may think that our posture and footprints are typical, but they're actually rather bizarre compared to those of most other animals. Digitigrade theropod dinosaurs nearly always held their metatarsi almost vertically, elevating their ankles well off the ground, so their footprints typically have only toe impressions. But a few *Grallator* and *Eubrontes* tracks at the SGDS have "heel" impressions made by the metatarsus (fig. 3.12). Simply sinking into loose sediment could bring the metatarsals into contact with the ground and leave an impression, though such tracks

3.14. *A–B, Characichnos* tracks (*A*, SGDS 167; *B*, SGDS SW103) made by small theropods while swimming. These theropods, which would have made *Grallator* tracks while on land, paddled with their feet to swim. In the shallowest water, more of the foot contacted the sediment for a longer period, creating long swim tracks (*A*); as the water deepened, less of the foot contacted the sediment for a shorter period, creating shorter swim tracks (*B*). Quarters for scale.

would be especially deep. In some cases, tracks with metatarsal impressions indicate that the track makers crouched down and rested on their metatarsals, in the same way that large birds do when they crouch or sit. In other cases, on slippery surfaces, the animals may have crouched (and widened their stances) just to stabilize themselves against slipping. In rare instances, the "heel" impressions indicate that the track maker slipped in the wet sediment and skidded, trying to regain its balance (sadly, no known trace records a dinosaur falling over). When the metatarsus was lowered to the ground, the tiny first toe would also have made an impression—these are instances of *Eubrontes* that have impressions of the first toe,

like *Gigandipus* tracks. But *Gigandipus* tracks do not typically have metatarsus impressions. They therefore may have been made by different theropods—ones with longer first toes— than those that made *Eubrontes* tracks. Or, as mentioned earlier, they may have been made by digitigrade theropods whose feet sank more deeply into the sediment, allowing the first toe to contact the ground but without a horizontal, trace-making metatarsus.

A few *Eubrontes* trackways include short grooves made when the track makers' tails occasionally touched the ground. Based on the ways theropod tail vertebrae (backbones) fit together, theropod tails were held straight out behind the rest of their horizontal bodies, so

3.14. *C*, The small theropod *Megapnosaurus* transitioning from walking on land to swimming in Lake Dixie. On land, the dinosaur made normal *Grallator* tracks, but as it began to float in the water, it swam by paddling with its hind limbs and feet, and the raking of its toe tips through the sediment created long, and then short, *Characichnos* tracks instead of *Grallator* tracks. (*Photo A by Anna Oakden; B by Jerry D. Harris; restoration by H. Kyoht Luterman.*)

how the tails dragged on the ground in these few track makers is unclear. Perhaps the animals were injured, with parts of their tails broken and sagging. Alternatively, these track makers could have engaged in an unusual behavior that tilted their bodies away from horizontal and brought the tips of the tails downward. However, we know from both the rarity of tail-drag traces and the anatomies of dinosaur tail bones that old-style pictures of bipedal dinosaurs with

kangaroo-like, vertically oriented bodies and curved tails dragging on the ground were impossible. Even the bipedal dinosaurs had horizontal bodies, the same as the quadrupedal ones, a very different posture than that of the vertically oriented human body.

Several SGDS *Grallator* tracks occur in trackways with exceptionally long step lengths (the distances between successive right or left footprints). In fact, the smallest tracks yet found at the SGDS, which are only 0.8 in. (2 cm) long, have step lengths ranging from 9 to 14.2 in. (23 to 36 cm). Another set of tracks, with footprint lengths of 2 in. (5 cm), has a step length of 22.4 in. (57 cm). A longer step length indicates that the track maker was moving faster, probably running (fig. 3.13). When running, the body leaves the ground completely and travels through the air for short periods before landing (this is actually the definition of running; running has nothing to do with speed), so the distances between steps are longer. Most of the theropod tracks at the SGDS, however, have short step lengths, indicating slower, probably walking gaits. This makes sense because most animals, including humans, spend far more time walking than they do running.

Swimming

Walking was probably the most commonly used gait (it conserves the most energy); running was less frequent. But these were not the only possible modes of locomotion. Thus far, we have not mentioned one of the most unusual track types found at the SGDS. This track type, called *Characichnos*, was made by swimming animals—at the SGDS, mostly by theropod dinosaurs. *Characichnos* traces usually consist of three (sometimes two or only one) long, parallel scrape marks. In the *Characichnos* traces at the SGDS, the middle mark is longer and deeper than the outer two marks (figs. 3.14*A*, *B*). These scrape marks were made by the claws at the tips of the toes of tridactyl theropods as they paddled their feet, duck-like, while swimming in Lake Dixie—yet more evidence that theropods went into bodies of water and were capable of swimming (fig. 3.14*C*). The SGDS has the world's largest and best-preserved collection of dinosaur swim tracks. Previously, a few possible

theropod swim tracks had been found in a few places around the world, but paleontologists debated whether they were truly made by theropods, which, bizarrely, were once thought to be afraid of water. The SGDS specimens settled the debate once and for all. Most of the *Characichnos* traces are *Grallator* size, and in some cases they occur in the same spot as nonswimming *Eubrontes* tracks, indicating that they were made in a place in Lake Dixie where the water was deep enough for the small, normally *Grallator*-making theropods to float and swim, but not so deep that it prevented the larger *Eubrontes* track makers from walking normally. A few, larger *Characichnos* traces are *Eubrontes* size and were made in deeper water, where even the larger *Eubrontes* track makers were buoyed in the water and had to swim.

In one place on the Top Surface Layer, tracks in a *Batrachopus* trackway along the top of a ridge change from typical, pentadactyl *Batrachopus* tracks into tiny swim tracks as the track maker moved down a slope and entered water pooled at the edge of Lake Dixie. The swimming portion of the trackway would also constitute *Characichnos* tracks, though of a different sort (a different ichnospecies) than the theropod ones: these tracks have four scrape marks, not three, because some early crocodylians, such as *Protosuchus*, had only four toes on their hind feet. Trackways such as this, which change from one type of track to another type along a trackway, demonstrate that the same animal can make different kinds of traces, depending entirely on the surrounding environmental conditions.

The Crouching Dinosaur

Finally, one of the rarest and most unusual traces preserved at the SGDS is a *Eubrontes*

3.15. *A*, Photograph, and *B*, schematic diagram of unusual *Eubrontes* tracks (SGDS 18.T1, on the Top Surface) made by a large theropod that crouched in the sediment, leaving not only footprints but also heel impressions, handprints, hip impressions, and tail traces. First the dinosaur sat, creating impressions of the ischium (hip bone), hands, and feet with metatarsal impressions. Second, it scooted forward a little before settling again, creating (third) new foot and ischial impressions. After some time (fourth), it stood and walked away, creating typical *Eubrontes* impressions. The hand impressions show that the fingers pointed inward, and the heel impressions show that theropods sat in exactly the same way that modern birds do. One foot (30 cm) ruler for scale. *C*, The large theropod dinosaur *Dilophosaurus* in the posture necessary to create the crouching *Eubrontes* tracks. *(Photo by Jerry D. Harris; restoration by H. Kyoht Luterman.)*

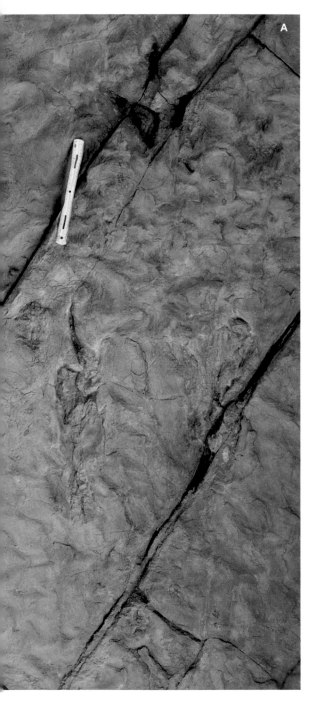
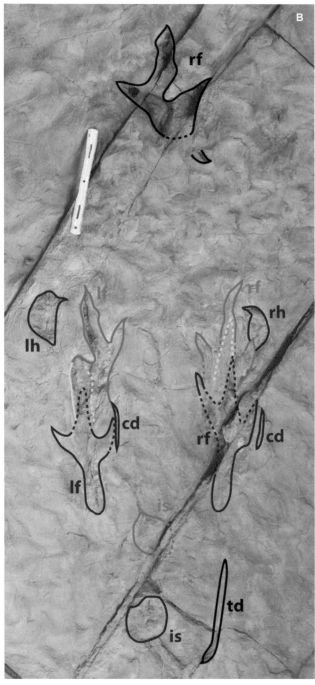

━━ first

━━ second

━━ third

━━ fourth

cd = claw drag
is = ischium
lf = left foot
lh = left hand
rf = right foot
rh = right hand
td = tail drag

trackway 73.2 ft. (22.3 m) long on the Top Surface (fig. 2.2), now preserved within the SGDS museum. Near the beginning of this trackway are crouching marks made when the track-making theropod emerged from Lake Dixie, sat down on the sediment, and then shuffled forward and crouched a second time, creating two overlapping crouching impressions (fig. 3.15). As the theropod rested on the ground, its downward-pointing rear hip bones (ischia) also made impressions between and behind the "heels" of the footprints; the end of its tail also

c

touched the ground, creating a narrow groove far behind the crouching traces. Why this large theropod sat down on this spot is anybody's guess, but we are fortunate that it chose a good spot for making such unusual trace fossils. Only six other traces made by such crouching theropods are known anywhere else in the world, and the SGDS specimen is one of only three on exhibit (the others are in remote places around the world). But the SGDS one is the best preserved and has many features lacking in all the others.

This unique trace fossil is important because of the information it preserves about early theropod dinosaur behavior and the relationship of these early theropods to modern birds. First, the fossil's foot impressions possess "heel" marks, made when the theropod crouched in a manner identical to how large, modern birds crouch. More important is that the trace includes impressions of the hands, an extremely rare phenomenon for a bipedal animal like a theropod. What the handprints say about how the theropod held its hands is also unusual and unexpected. The fingers and toes of most quadrupedal animals point forward when the palms of the hands and soles of the feet are on the ground. In a typical relaxed, bipedal posture, the hands of humans dangle with the palms facing inward, toward each other, and the fingers pointing down. But when humans get down in a quadrupedal posture (on all fours), they imitate quadrupedal animals and put the palms of their hands on the ground with their fingers pointing forward. Unlike most other animals, though, humans have the ability to turn their hands palms up or palms down; most quadrupedal animals have their hands permanently "locked" in the palms-down position; if they could stand up bipedally, their palms would face backward. Humans are able to turn their hands palms up and palms down because of unusual adaptations in their lower arm bones: the elbow end of one bone (the radius) rolls around the other (the ulna). Very few other animals have this ability, and theropod dinosaurs lacked it—as bipeds, theropods had no need for an ability to rotate their hands to face palms down. Instead, their hands could face only one way: with the palms facing inward, toward one another, permanently ready to grasp prey. (Think of it this way: if basketballs had existed in the Mesozoic, theropods could have held a basketball but could not have dribbled it.) Birds, which evolved from small theropods in the Late Jurassic, still have this limitation. Paleontologists already knew from skeletal fossils that later theropods had this limitation, but no one knew how far back in time, and how far down the theropod evolutionary tree, it went. The crouching traces at the SGDS show that even some of the earliest theropods had similar limitations—trace fossils to the rescue!

After the crouching theropod stood back up, it began walking across the undulating Top Surface, creating more typical *Eubrontes* tracks. When the animal went up the sides of some of the ridges, the very tip of its tail occasionally dipped down and touched the surface, leaving small, short tail drag traces. Where it went, beyond the preserved end of the trackway, is a mystery. A few of its left footprints also have tiny divots unattached to the rest of the prints, made by the claws at the ends of its dangling first toes—unusual for a *Eubrontes* track maker.

Highlight: What Is a Fossil?

Fossils are the remains and traces of ancient life. To define "ancient," paleontologists usually draw an arbitrary line at ten thousand years old: any remains of living things older than that are fossils, while any younger remains are subfossils. By "life" is meant *any* living organism. Many people think that the words "fossil" and "dinosaur" mean the same thing, but this is incorrect—not all fossils are of dinosaurs. There are fossil mammals, fossil fishes, fossil clams, fossil insects, fossil plants, even fossil bacteria. If it was alive, and its remains or traces are preserved in rocks, then it is a fossil. Only a very few fossils are of dinosaurs.

Many people also think that the words "fossil" and "extinct" mean the same thing, but this too is incorrect: there are many species that have representatives that are alive and well today that also lived long ago; this is particularly true of many plants and invertebrates. For example, there are fossil horsetails (a kind of plant) that are several hundred million years old that are identical to living horsetails. Just because an organism occurs as a fossil does not necessarily mean that the species to which that organism belongs is extinct.

Exactly how and why fossils form is not well understood—paleontologists have only recently begun to examine this process in detail. However, some things are fairly certain. To become a fossil, the remains or traces of an organism must be buried by sediment before they can be destroyed at the surface. Body parts must be buried before they are eaten by scavengers or broken down entirely by fungi and bacteria; traces, such as footprints, must be buried before natural weathering processes erase them. For both body parts and traces, newer research indicates that particular kinds of bacteria must be present that actually protect the remains and aid in the fossilization process. But even burial is no guarantee of becoming a fossil: many underground conditions can also destroy fossils, so there are very specific burial conditions that also must be met. Because these conditions are rare, not everything becomes a fossil—if everything that ever lived became a fossil, then the surface of the Earth would be covered with thousands of feet of dead bodies.

The odds of something becoming a fossil cannot be accurately calculated because we do not yet fully understand all the factors involved, but fossilization seems to be a rare event except in some special environments. However, the fossilization process does not necessarily take millions of years. In some special conditions, fossils can form even in a matter of days.

BODY FOSSILS

CHAPTER 4

As if the phenomenal tracks and traces weren't enough by themselves, the SGDS is also one of very few places in the world where trace fossils and body fossils occur together. (Body fossils, remember, are the actual remains of organisms.) Typically, the processes that make the preservation of tracks possible tend to destroy body fossils, and vice versa. But both

4.1. Stromatolite fossils (SGDS 669) made by colonies of cyanobacteria living on the bottom of Lake Dixie. The small, dark red blotches are iron concretions that contain fish scales or bones that were deposited on top of or within the stromatolites. Penny for scale. *(Photo by Anna Oakden.)*

types of preservation conditions occurred in the Whitmore Point Member of the Moenave Formation, albeit in different layers: most (but not all) of the tracks are in relatively low strata, and most of the body fossils are in higher strata (fig. 2.2). But all were deposited within a few hundreds of thousands or millions of years, a short enough span of time, geologically speaking, to allow paleontologists to assume that the same species of plants and animals lived in and around Lake Dixie at any point during that time.

Invertebrate Body Fossils

In some of the offshore (underwater) deposits from Lake Dixie, large patches of rock have been collected that resemble low, irregular lumps with a rough, slightly knobby texture (fig. 4.1). These unassuming structures are actually fossils called stromatolites. Stromatolites are mounds that form when layers of certain photosynthetic bacteria called cyanobacteria get covered by sediment and grow new layers. Even today, cyanobacteria coat the bottoms of lakes and parts of oceans—if you have ever stepped into a lake or pond and felt a coating of slime on the sediment at the bottom, that was a film of harmless cyanobacteria (and probably some algae, too). In some ocean environments,

4.2. Fossil ostracod (O) and conchostracan (C) shells (SGDS 755) from the SGDS. Penny for scale. *(Photo by Jerry D. Harris.)*

fields of stromatolites that can grow to be several meters across and a few meters tall cover wide areas. Lake stromatolites, however, rarely get this big, and in Lake Dixie, they were apparently restricted to small mounds—the biggest ones known from the Whitmore Point Member of the Moenave Formation are just 1 ft. (0.3 m) across, but most are much smaller. Various invertebrates and fishes feeding on them probably prevented them from getting any bigger; too much sediment washing into Lake Dixie also could have prevented growth by burying existing structures and forcing the cyanobacteria to start over on a new layer of sediment. Nevertheless, the stromatolites show that Lake Dixie was not much different from most modern lakes in hosting thriving colonies of bacteria.

Some of the common invertebrate body fossils found at the SGDS are the almost microscopic shells of ostracods and conchostracans. Because both of these organisms have paired shells, they are sometimes mistaken for tiny clams (fig. 4.2), but neither was actually a clam: both were

crustaceans, related to crabs and lobsters, and both are common in larger lakes today. Inside (and attached to) each pair of ostracod or conchostracan shells was a tiny animal that looked vaguely like a shrimp or a flea. Fossil ostracod and conchostracan shells are often useful for determining very specifically the ages of the strata in which they are found (see "Highlight: How Do We Know How Old Rocks and Fossils Are?" at the end of chapter 2), but the ones found so far at the SGDS are not well enough preserved for this purpose. However, they are an interesting addition to the site's list of animals because trace fossils made by ostracods and conchostracans are not yet known with certainty there, so without these shells, we may not have known they were present at the site. (The Lake Dixie ostracods are too small to have made the *Lockeia* traces known from the site, but at least some *Lockeia* traces may have been made by conchostracans.) Similarly, many of the invertebrate trace fossils at the SGDS were made by worms, insects, spiders, horseshoe crabs, and possibly clams, for which we do not

4.3. *A–B,* Dorsal fin spine (*A;* SGDS 828) and jaw fragment full of blunt, rice-grain-sized teeth (*B;* SGDS 857) of the hybodont shark *Lissodus johnsonorum* (*C*). Penny for scale. *(Photos by Anna Oakden; restoration by H. Kyoht Luterman.)*

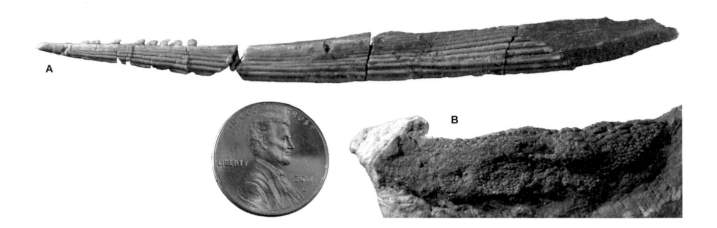

(yet) have body fossils from the site. The body fossil and trace fossil records *both* help fill in our picture of the Early Jurassic ecosystem preserved at the SGDS.

Vertebrate Body Fossils
Fishes

Many fish fossils—bones, teeth, and scales—have been found at and near the SGDS. At least some of these fishes were probably the makers of *Undichna* and *Parundichna* traces. Some of the fossil bones belong to a previously unknown species of hybodont shark. These sharks are extinct today but were common in the Mesozoic. Unlike most living sharks, many hybodonts lived in freshwater. Shark skeletons are made of cartilage, not bone, so the most common shark fossils are teeth because those were the only hard, mineralized parts of their bodies. Hybodonts, however, also had a few other solid bones: most noticeably, they had sharp, serrated, bony spines in front of their dorsal (back) fins (fig. 4.3*A*), somewhat like those of modern bullhead and horn sharks. Hybodont fossils found at the SGDS include several of these dorsal fin spines, as well as some tiny teeth the size and shape of rice grains (fig. 4.3*B*). Most of the hybodont fossils from the SGDS come from the upper Whitmore Point Member (fig. 2.2), above the Johnson Farm Sandstone Bed, but they were certainly also present in Lake Dixie at the time the Johnson Farm Sandstone Bed and its footprints formed. The shapes of the SGDS teeth and fin spines are similar to those of other species of a hybodont called *Lissodus*, but they were not quite the same as those of any known species. They were something new, so they were placed in a new species: *Lissodus johnsonorum* (in honor of Dr. Johnson and his wife, LaVerna). *Lissodus*

johnsonorum was hardly a terrifying shark: it measured only a little over 1 m (3.3 ft.) in length (fig. 4.3*C*), and its tiny, blunt teeth were good for crushing shells of ostracods, conchostracans, clams, and other small invertebrates.

All the other fish fossils found so far at the SGDS are of bony fishes, not cartilaginous fishes like sharks. In terms of overall body shape, these bony fishes would have looked very similar to most living bony fishes, but they differ in many important anatomical details of the skull and body. Most of the SGDS fish fossils belong to groups of bony fishes that are now extinct. At least two specimens belong to a group of fishes called palaeoniscoids (fig. 4.4*A*, *B*); these fossils are currently being studied. Like the hybodont shark, they will also likely belong to one or more new species because no other Early Jurassic palaeoniscoids are known in the southwestern United States, though some are known from the underlying Chinle Formation.

By a wide margin, the most common fish fossils at the SGDS belong to a group called semionotids. Semionotid bodies were covered in thick, diamond-shaped, enameled scales, similar to those on distantly related modern-day gars (fig. 4.4*C*). These armor-like scales provided some measure of protection from predators. Semionotids are relatively short from head to tail, but somewhat tall from belly to back. At first, all specimens found at the SGDS were thought to belong in the genus *Semionotus*. Several species of *Semionotus* are well known from Late Triassic–age rocks in the eastern United States; specimens found in the Chinle Formation of southeastern Utah in the 1950s and 1960s were also attributed to *Semionotus*. One species, named *Semionotus kanabensis* in 1950, was based on specimens found in the Moenave Formation in the 1880s near Kanab,

Utah. However, detailed study of the specimens from both the Chinle and Moenave Formations by former SGDS intern Sarah Gibson revealed numerous subtle but important features that differ from those of *Semionotus* specimens from the eastern United States. These differences were enough to warrant separating them from *Semionotus* into a new, but related, genus called *Lophionotus*. The specimens from near Kanab

therefore belong to *Lophionotus kanabensis*. Many of the SGDS semionotid fossils likely also belong to this species, but others may belong to different species of *Lophionotus* (fig. 4.4D), or possibly even another new, as yet unnamed genus.

Lungfish are peculiar, vaguely eel-like freshwater fishes. The three types that survive today are rare: one (endangered) genus lives

4.4. *A–B,* Part of the body, including a large patch of scales (*A*; SGDS 1241), of an unnamed palaeoniscoid fish (*B*). In *A*, the head is to the right and the tail to the left. *C–D,* A mostly complete skeleton and scales (*C*; SGDS 894) of a semionotid fish, probably a species of *Lophionotus* (*D*). In *C*, the head is to the lower left and the tail to the right. Pennies for scale. *(Photos by Anna Oakden; restorations by H. Kyoht Luterman.)*

in Australia, one lives in South America, and one lives in Africa. In the Mesozoic, however, relatives of modern lungfish were very common. Instead of lots of individual teeth, lungfish have upper and lower pairs of thick, heavy tooth plates; these fossilize better than their other, lightweight bones. SGDS volunteer Sally Stephenson found a lungfish tooth plate (fig. 4.5A) in 2004 in the Whitmore Point Member of the Moenave Formation near St. George. Even though it was not found at the SGDS site, it was in strata deposited in the same Lake Dixie, so this lungfish certainly also lived at and near the site. This tooth plate is similar to tooth plates of the now-extinct lungfish *Ceratodus* but differs in many details from those of all known species. As a result, this lungfish was placed in a new species, *Ceratodus stewarti*, in honor of Darcy Stewart, a major contributor to the SGDS (fig. 4.5B). This tooth plate is only the second ever found in Early Jurassic–age rocks of the American Southwest (the other is from the Kayenta Formation in Arizona) and is one of the oldest *Ceratodus* fossils in North America.

The biggest fish known from Lake Dixie was a coelacanth. Today coelacanths are rare and live in very deep waters of the Indian Ocean. In the Mesozoic, however, coelacanths were much more common, and many lived in freshwater. The SGDS coelacanth is known from several large skull and jaw bones (fig. 4.5C) and a partial

tail from near the top of the Whitmore Point Member of the Moenave Formation. These bones belonged to individuals around 4 ft. (1.2 m) long. The SGDS coelacanth is by far the largest Early Jurassic coelacanth anywhere in the world, and most likely the largest known freshwater coelacanth on record. (It is dwarfed, however, by some of its ocean-going relatives from the Cretaceous Period.) No other Early Jurassic coelacanths have been found in the southwestern United States, though coelacanths called *Chinlea* and *Quayia* are known from the Late Triassic–age Chinle Formation in Colorado, New Mexico, and southeastern Utah, as well as strata of equivalent age in Texas. The SGDS specimens are similar to *Chinlea* (fig. 4.5*D*) but almost certainly belong to a new species.

Dinosaurs

Because theropod tracks such as *Grallator* and *Eubrontes* are so abundant at the SGDS, we know theropods lived in the area. Nevertheless, finding theropod body fossils to complement the tracks was a wonderful surprise. Theropod bones are common in only a very few places in the world, but they are particularly rare in the Moenave Formation. The SGDS theropod bones actually come from a layer of the Whitmore Point Member of the Moenave Formation that is several feet (a few meters) above, and therefore

A

4.5. *A–B*, Tooth plate (*A*; Utah Museum of Natural History VP 16027) of the lungfish *Ceratodus stewarti* (*B*). *C–D*, Lower jaw bone (*C*; SGDS 892) of the unnamed coelacanth from the SGDS (*D*). Pennies for scale. *(Photos by Anna Oakden; restorations by H. Kyoht Luterman.)*

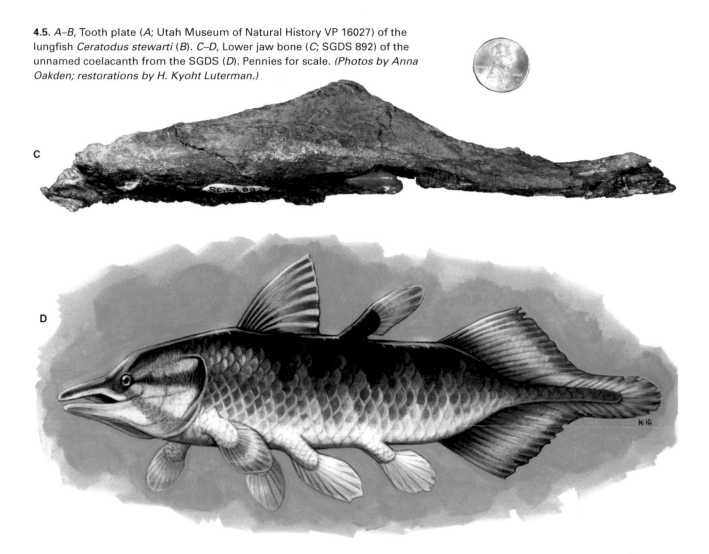

younger than, the track-bearing layers (fig. 2.2). Still, these bones likely belong to individuals of the same species that made at least some of the SGDS tracks.

The most common theropod body fossils found thus far are teeth (fig. 4.6*A*, *B*). Because most theropods were carnivores, theropod teeth are fairly simple: usually rather flat from side to side and curved slightly backward. In most theropods, the front and back edges of the teeth were lined with tiny serrations, somewhat like those on a steak knife. (Contrary to popular depiction, theropod teeth were not "sharp as knives," but the serrations on the edges did help their teeth cut through meat efficiently.) Theropods, like all other dinosaurs (as well as crocodylians), each had an endless supply of replacement teeth, so when old teeth were worn, they simply fell out and were replaced by new ones. The old shed teeth are therefore the most commonly found theropod fossils, especially in places where the theropods were feeding. At least one of the teeth found (fig. 4.6*A*) may belong to *Megapnosaurus*, but this is difficult to say for certain because, as with footprints, many different theropods had virtually identical teeth.

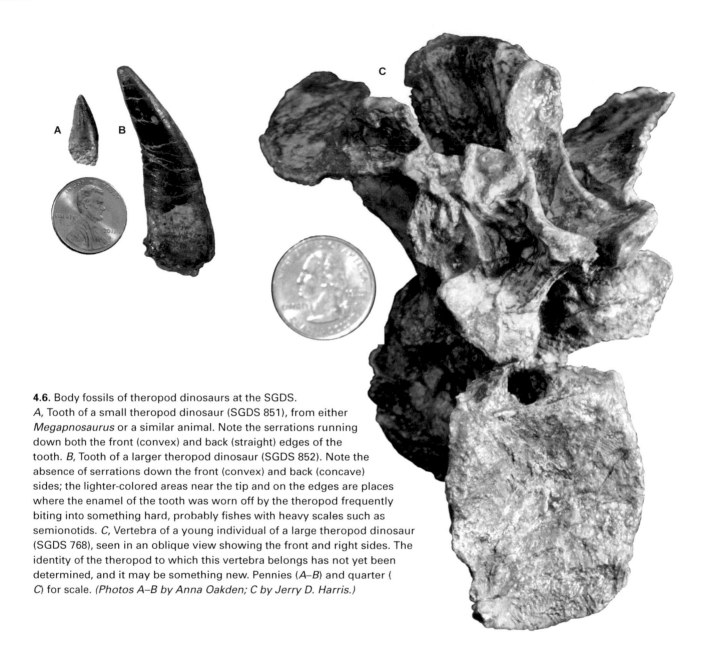

4.6. Body fossils of theropod dinosaurs at the SGDS. *A*, Tooth of a small theropod dinosaur (SGDS 851), from either *Megapnosaurus* or a similar animal. Note the serrations running down both the front (convex) and back (straight) edges of the tooth. *B*, Tooth of a larger theropod dinosaur (SGDS 852). Note the absence of serrations down the front (convex) and back (concave) sides; the lighter-colored areas near the tip and on the edges are places where the enamel of the tooth was worn off by the theropod frequently biting into something hard, probably fishes with heavy scales such as semionotids. *C*, Vertebra of a young individual of a large theropod dinosaur (SGDS 768), seen in an oblique view showing the front and right sides. The identity of the theropod to which this vertebra belongs has not yet been determined, and it may be something new. Pennies (*A–B*) and quarter (*C*) for scale. *(Photos A–B by Anna Oakden; C by Jerry D. Harris.)*

Even if the tooth does belong to *Megapnosaurus*, it still would not mean that *Megapnosaurus* necessarily made *all* the *Grallator* tracks at the SGDS, but it would be a good indication that it made at least some of them.

Some much larger teeth, from a larger theropod, are rather unusual: instead of being flat from side to side like typical theropod teeth, they are fatter and more conical (fig. 4.6*B*). These teeth are good for grabbing and holding on to slippery fishes; serrated, flatter teeth would just shred a fish before it could be swallowed. The enamel on the tips and edges of the larger SGDS teeth, as well as any serrations they may have had, has been worn off in a few places, indicating that the teeth often came in contact with hard, tough food. A few theropods, most famously *Spinosaurus*, have similar teeth, and they are thought to have been piscivorous, but none of these are from the Early Jurassic epoch—most, including *Spinosaurus*, lived in the Early and Late Cretaceous, 60–100 million years later. So the peculiar Moenave Formation teeth may belong to a new kind of theropod dinosaur and might show that piscivory evolved more than once among theropods. More bones of this animal must be discovered before it can be named, however. The close association of these teeth with the *Eubrontes*, *Gigandipus*, and *Kayentapus* tracks at and around the SGDS may mean that this unknown theropod is a better candidate than *Dilophosaurus* for the maker of these tracks. This highlights the danger of saying that *Dilophosaurus* must have made *Eubrontes*, *Gigandipus*, or *Kayentapus* tracks in the Moenave Formation just because nothing else is yet known from that time in the area.

Unlike teeth, other bones were not continually replaced throughout a theropod's life. So one of the biggest surprises yet discovered at the SGDS is a single vertebra (fig. 4.6C) from the middle of a theropod's back. The vertebra came from the same layer as the teeth. The top part of the vertebra is not fused to the bottom part, indicating that the animal was not a full-grown adult when it died. This specimen is currently being studied to determine whether it belongs to *Megapnosaurus*, *Dilophosaurus*, or something completely new, perhaps even the theropod that had the conical, fish-eating teeth.

Plant Body Fossils

In any land-based ecosystem, plants form the base of the food web, and that was certainly true around Lake Dixie in the Early Jurassic. Plants are usually much more abundant than animals,

and they continually shed and regrow leaves and branches. Plant parts are therefore easily and often buried in stream and lake sediments. Special conditions are needed to preserve plant fossils, however, and thus far only bits and pieces of fossil plants have been found at the SGDS. But these fragments are important because they enable paleontologists to reconstruct the flora that formed the base of the food web in the Lake Dixie area. Several layers in the upper part of the Dinosaur Canyon Member of the Moenave Formation in St. George, a little below the Johnson Farm Sandstone Bed, have produced plant fossils. A few plant impressions are also known from the Whitmore Point Member: they were found along with the tracks on the Top Surface layers of the Johnson Farm Sandstone Bed within the SGDS museum. So far, at least seven different kinds of plant have been identified, and doubtless there were many more around in the Early Jurassic.

Several impressions of horsetails (fig. 4.7A) were found on the Top Surface of the Johnson Farm Sandstone Bed. These belong to the genus *Equisetum*, the same genus of horsetail that is alive today (fig. 4.7B). Four incomplete leaves (fig. 4.7C) belong to an extinct kind of fern called *Clathropteris* (fig. 4.7D), but it is not yet known whether these leaves belong to a new species. Horsetails and ferns prefer wet environments, which would have existed around Lake Dixie and the rivers that fed it in the Early Jurassic.

The most common SGDS plant fossils are of conifers—early, distant relatives of today's pine, fir, spruce, and juniper trees. These fossils consist mainly of branches and twigs, some of which still bear short leaves and even small cones along the branches. Fossils of five distinct types of conifer have been found. One belongs to a species of the extinct conifer *Pagiophyllum* (fig. 4.8*A*, *B*), and another (a small leaf fragment) to a species of *Podozamites* (fig. 4.8*C*). Two belong to new species, never before discovered. One of these (fig. 4.8*D*, *E*) was named *Saintgeorgia* (after the city of St. George); the other

(fig. 4.8*F*, *G*) was named *Milnerites*. Some of the *Saintgeorgia* specimens are spectacular because they preserve branches with both cones and shoots still attached (fig. 4.8*F*). Parts of fossil cones have also been found that belong to the extinct conifer *Araucarites*

D

4.7. Horsetail and fern fossils from the SGDS. *A*, Partial fossil of the horsetail *Equisetum* (SGDS 569). *B*, Example of a modern, living *Equisetum* plant on the Isle of Wight, England. *C–D*, Partial leaf fossil (*C*; SGDS 503 and SGDS 512A) and restoration of the leaf and whole plant of *Clathropteris* (*D*). Pennies for scale. *(Photos A, C by Anna Oakden; B by Jerry D. Harris; restoration by H. Kyoht Luterman.)*

(fig. 4.8*H*). These conifers were likely among the tallest trees around Lake Dixie.

Plant fossils from the SGDS and surrounding area were probably washed by streams into debris piles on mudflats, or into topographic depressions in and around Lake Dixie. To date,

no other localities containing identifiable plants from the Early Jurassic have been documented in western North America, making these specimens and localities very important for understanding how plants, and whole ecosystems, evolved from the Triassic into the Jurassic.

4.8. Conifer fossils at the SGDS. *A–B*, Branch (*A*; SGDS 491) and restoration of the leaf and whole plant (*B*) of *Pagiophyllum*. The circular structures surrounding the branch are natural casts of raindrop impressions. *C*, Restoration of the leaf and whole plant of *Podozamites*. *D–E*, Branch with leaves and cones (*D*; SGDS 517B) and restoration of the leaves, cones, and whole plant (*E*) of *Saintgeorgia*. *F–G*, Branch with leaves (*F*; SGDS 513A) and restoration of the leaves and whole plant (*G*) of *Milnerites*. *H*, Isolated cone scale (upper right, SGDS 555) and whole cone (left, SGDS 556B; lower right, SGDS 556A) of *Araucarites*. Pennies for scale. *(Photos by Anna Oakden; restorations by H. Kyoht Luterman.)*

LAKE DIXIE

CHAPTER 5

All the fossils at the SGDS are, by themselves, certainly fascinating and important, but when combined with aspects of the site's geology, they paint an even bigger, more interesting picture. Imagine that someone traveled through time back to the beginning of the Jurassic and took a snapshot. Returning to the present, that person tore that snapshot into lots of little pieces and scattered them around southwestern Utah and northwestern Arizona. What we know now about Lake Dixie and its environment comes from assembling those little pieces. Some of those pieces have animals or parts of animals on them, some have parts of plants, and still others have bits of the ground or the lake. And we don't even have all the pieces yet! Nevertheless, "snapshots" of prehistory that can be reconstructed with as much detail as is preserved at the SGDS—that have as many pieces of the hypothetical photograph— are rare and highly prized by geologists and paleontologists because they provide much more detail than scientists usually have to work with when reconstructing the past.

We have already discussed the fossils and some aspects of the overall geology, but if we take a closer look at features on the rocks *around* the fossils, even more details come to light. These features, called sedimentary structures, are very important because they were created by the very environment in which all the Early Jurassic plants and animals lived. Most sedimentary structures are created by actions and interactions of wind, water, air, and sunlight. As with the tracks, geologists understand the formation of sedimentary structures preserved in sedimentary rock by observing the environmental conditions in which identical structures form today. Understanding the sedimentary structures allows us to put the plants and animals back into the proper environmental context in which they lived.

Sedimentary Structures

The most common and readily identifiable sedimentary structures on the Top Surface of the Johnson Farm Sandstone Bed—the track-covered surface visible as half of the floor inside the SGDS museum—are fossilized ripple marks (fig. 5.1*A*, *B*). These were made in loose sediment by wave action at the edge of Lake Dixie and by water moving in the streams that fed the lake. The ripples vary in size and shape from place to place, indicating different current speeds and directions. Ripples form most commonly in shallow water; in deep water, the movement of water at the surface doesn't often reach the bottom and cannot disturb the sediments. So ripples outline where Lake Dixie's shoreline was at the time the sediment that became the Top Surface was exposed. Most of the tracks on the Top Surface overlap and disrupt some ripples, indicating that the ripples were made first and the tracks slightly later (probably within days, if not hours). Ripples in other Whitmore Point Member strata indicate where there were underwater currents in the lake itself.

The Top Surface is not flat, and it is very complex. Most notably, the surface undulates: it has higher, hill-like ridges and lower, valley-like swales (fig. 5.1C). Powerful moving water in Lake Dixie scoured out low-lying swales along the shoreline, leaving behind ridges. Sediment later eroded down the slopes of the ridges, gradually filling the bottoms of some of the swales with new, thin layers of sediment. Some of the swales held small inlets of Lake Dixie, as indicated by particular kinds of ripple marks on the swale bottoms that formed when water flowed in and out along the same path. Ridges and swales also indicate the position of the shoreline of Lake Dixie, at least during the short phase of its existence when the Top Surface formed (remember, the position of the shoreline fluctuated over time). In a few places, rill marks are preserved on the slopes of the ridges; these formed when thin sheets of water flowed down the ridge slopes, carving frilly, shallow channels into the sediment (fig. 5.1D).

Some parts of the Top Surface also exhibit mud cracks (fig. 5.1E), created when the position of the shoreline of Lake Dixie changed. When the lake shrank, the shore extended in the direction of the lake, exposing previously submerged, wet sediment to sunlight and air, which dried it out. Alternatively, sediment wetted by rain during Early Jurassic monsoon seasons would have dried out when the dry season began. In either case, when wet sediment dries out, it tends to contract, creating a web of cracks in the surface. The Main Track Layer of the Johnson Farm Sandstone Bed—the one with all the natural casts of footprints now exposed on the overturned blocks in the museum, and the same one Dr. Johnson first discovered— also has many mud cracks (fig. 1.3B). However, these are natural casts of mud cracks: in the same way that the natural casts of footprints on the underside of the sandstone layer formed when sand filled in natural mold footprints on the muddy (now mudstone) layer underneath, the sand also filled in mud cracks on the muddy layer. The existence of mud cracks on different layers and in different places within the Whitmore Point Member indicates both the seasonal climate of the area and the fluctuation of the shoreline of Lake Dixie as it changed position over hundreds and thousands of years, just as modern lakeshores do.

Some places on the Main Track Layer surface and the Top Surface have tiny dots that resemble miniature impact craters, which is exactly what they are. These are raindrop impressions (fig. 4.8B), which form when drops of rain hit and make slight indentations in loose sediment on the surface. Like the footprints, these can be preserved only when the sediment they are in dries out and hardens a bit before being buried under newer sediment.

Raindrop impressions and mud cracks form under the air, and ripple marks form under shallow water, but there are also sedimentary structures that form under deeper water. Tool marks form when currents push objects, like small rocks or pieces of wood, through the sediment, creating grooves (fig. 5.1F). Like ripple marks, tool marks document the direction of water movement. Very large marks

5.1. Sedimentary structures at the SGDS. *A–B*, Ripple marks of different sizes and patterns, including long, continuous ripples (*A*) and short, curved, discontinuous ripples (*B*; both SGDS 18, on Top Surface), indicating different speeds and motions of moving water. The discontinuous ripples in *B* are crossed by three natural mold *Grallator* tracks proceeding from right to left. These tracks must have been made after the ripples formed because the tracks overprint and disrupt the ripples. *C*, Ridges and swales (SGDS 18) on the Top Surface. Small ripples and tracks made in wet, soupy sediment occur on the bottoms of some of the valley-like swales. Note the linear parallel joints (natural fractures) in *A–C*. *D*, Natural casts of rill marks (SGDS 347). *E*, Natural casts of mud cracks (SGDS 6). Note that the mud cracks are interrupted by the *Eubrontes* footprint in the lower left, rather than vice versa, indicating that the mud cracks formed first, and the theropod track was made later. *F*, Natural casts of tool marks (SGDS 621). *G*, Natural casts of flute marks (SGDS 249).One foot (30 cm) rulers (*A–B, E*), pennies (*D, F*), and quarter (*G*) for scale. *(Photos A–B, F–G by Jerry D. Harris; C–E by Anna Oakden.)*

that resemble tool marks, called flute marks (fig. 5.1G), form as masses of sediment avalanche down underwater slopes, creating deep gouges. Both tool and flute marks are preserved at the SGDS in parts of the Whitmore Point Member of the Moenave Formation and indicate submerged areas of Lake Dixie.

Fossils and Ancient Environments

One of the most striking features of the SGDS is the relationship between the distribution of the tracks and the sedimentary structures. Just as different sedimentary structures form above water and underwater, different tracks can be made only above water or underwater. *Characichnos* swim tracks, for example, could not have been made above water, and burrows made by insects or spiders could not have been made underwater (or the insects and spiders in them would have drowned). By mapping the exact positions of the

different track types and sedimentary structures, we can define the shoreline of Lake Dixie at the time the Top Surface was forming, and onshore (exposed to air) and offshore (underwater) areas become clear. At the SGDS, the shoreline had a roughly northeast–southwest trend that cut through the present-day northwest corner of the museum, though there were small inlets into some of the swales.

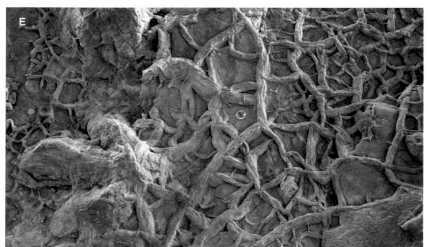

Many of the dinosaur trackways are either parallel to the lakeshore or at an angle to it; some are perpendicular to the shoreline. The latter may indicate places where theropods entered or left the lake, or walked along ridgetops. Unfortunately, the track-bearing layers are slightly tilted, as a result of tectonic shifts millions of years after the layers were deposited and buried. This tilt is such that, south of the museum, the track-bearing layers were eroded away by the Virgin River long before Sheldon Johnson began his excavations, and north of the museum, the layers angle farther underground, so we cannot see where the theropods were, or where they went. *Batrachopus* tracks seem to be concentrated on the ridgetops, indicating that the small, early crocodylians that made these tracks preferred walking across higher terrain, perhaps to better watch for hungry theropods.

Although none have been found in the Johnson Farm Sandstone Bed, many fish fossils have been recovered from higher (and therefore slightly younger) strata of the Whitmore Point Member of the Moenave Formation (fig. 2.2). At the site where the SGDS museum is now, where the Johnson Farm Sandstone Bed is exposed, these layers once composed the upper portion of the hill that Dr. Johnson removed before discovering the tracks. These fish-bearing layers represent a time, perhaps just a few hundreds of thousands of years after the Top Surface was formed, when the same site was under deeper water—again, Lake Dixie's margins changed position over time. Recently, fish scales have been found in the mudstones that were deposited directly on top of the Top Surface layers, indicating that Lake Dixie's water level rose, and the margins of the lake expanded, not long after the Top Surface tracks were made.

Filling In the Lake Dixie "Snapshot"

Added together, the fossils and sedimentary structures paint a very detailed picture of a tiny piece of the world roughly 200 million years ago. A seasonally wet, warm climate lay over a vast lake fed by several streams. The lake teemed with a wide variety of fishes, most commonly heavily armored semionotids and palaeoniscoids, but also small, spine-finned hybodont sharks, larger lungfish, and still-larger coelacanths. On the lake bottom, among patches of bacterial stromatolites, were millions of tiny ostracods, conchostracans, and perhaps clams and snails. Scuttling among the stromatolites were horseshoe crabs. Worms burrowed in the sediment below the water.

Around the shores of the lake, and along the banks of the streams that fed it, stood groves of conifers surrounded by an understory of shorter horsetails and ferns. Beetles, other insects, and spiders thrived among the plants. The barren sediment on the lake's beach was frequently crossed by small and large theropod dinosaurs and early crocodylians going into the lake to bathe and to feast on the abundance of fishes, invertebrates, and, occasionally, each other. Small herbivorous ornithischian dinosaurs, lizard-like sphenodontians, and fuzzy "protomammals" would occasionally dart from the cover of the conifer groves and shrubby ferns, cross the beach, and grab a quick drink or bath before being spotted by a hungry theropod.

We know that this picture, although very detailed, is not complete. First, there is no reason to think that all the plants and animals we have already discussed were the *only* representatives of their groups that were around. Second, and more importantly, we know that many other kinds of plants and animals were alive during this early part of the Jurassic Period. Many of these other plants and animals are known from multiple, often far-flung places around the world. Therefore, there cannot have been major geographic barriers preventing other animals from expanding into what is now the American Southwest during the Early Jurassic. So even though no fossil evidence of them has yet been found at the SGDS, we can hypothesize that they were there—future discoveries at and around the SGDS may yet turn up hard evidence.

Among plants, there would have been many species of ferns and conifers other than *Clathropteris*, *Pagiophyllum*, *Saintgeorgia*, and *Milnerites*. Very likely there would also have been other kinds of plants, including cone-bearing cycads and now-extinct bennettitaleans, which had leaves and bodies similar to those of cycads but also large, flower-like structures. Many cycads and bennettitaleans would have looked like giant pineapples sitting on the ground; others looked vaguely like small palm trees. There may also have been early relatives of modern ginkgo trees, and now-extinct plants called seed ferns, which were trees that had fernlike leaves but that reproduced with large seeds instead of microscopic spores. An enigmatic plant called *Sanmiguelia* has been found in the Whitmore Point Member in Zion National Park; whatever it was, it probably lived around Lake Dixie, too. There would not, however, have been colorful flowers or fruit around Lake Dixie—flowering plants did not evolve until the early Cretaceous Period, more than 60 million years later. There were also no grasses, which are a type of flowering plant; the oldest known grasses come from the Cretaceous. The waters of Lake Dixie would have supported many different kinds of algae, too.

Among invertebrates, there would have been flies, dragonflies, mayflies, cockroaches, and crickets, but many other familiar insects, such as ants, butterflies, and bees, had not yet evolved. Centipedes, millipedes, spiders, and scorpions were also present. In the water, there were probably also crayfish, snails, and lots of other small aquatic invertebrates similar to those that might be found in today's lakes and ponds.

Among vertebrates, doubtless many other kinds of fishes than the few we now know about lived in the lake. The oldest, most primitive frogs known come from the Kayenta Formation of Arizona, and it is possible that their earlier relatives lived a few million years earlier in Lake Dixie. Salamanders are close relatives of frogs, so if frogs lived in and around Lake Dixie, salamanders were almost certainly present, too. Turtles had evolved earlier in the Triassic and are known from the Chinle Formation in New Mexico, as well as the Kayenta Formation in Arizona, so primitive turtles likely also basked along the lakeshore. Along with the lizard-like sphenodontians, there might have been a few very primitive true lizards, but no snakes (which did not evolve until later in the Jurassic).

The skies above Lake Dixie would have been home to some small, early, long-tailed pterosaurs, the flying reptiles often incorrectly called pterodactyls. These would have fed on insects, fishes, and small vertebrates. Pterosaurs make highly distinctive tracks when they land on the ground, but none have yet been found at the SGDS or anywhere else in the world in Early Jurassic–age rock. But Early Jurassic pterosaur bones and skeletons are known in several places around the world, including western North America, so they were certainly present. Fossils that belong to the long-tailed pterosaur *Dimorphodon* are known from rocks in England and Mexico that are slightly younger than the Moenave Formation, and a similar pterosaur called *Rhamphinion* is known from the back portion of a skull from the Kayenta Formation in Arizona.

Big, plant-eating dinosaurian relatives of theropods, called prosauropods, must also have been present because their bones are found in Early Jurassic–age rocks around the world. The feet of these huge, long-necked, small-headed dinosaurs would have made very distinctive footprints (though some workers have proposed that prosauropods were the real makers of *Eubrontes* tracks), but neither tracks nor bones of these animals have yet been found in the St. George area or in the Moenave Formation.

However, possible prosauropod teeth have been found in Texas in strata equivalent to the Chinle Formation; similarly, Late Triassic prosauropod tracks are known from both the American Southwest and East Coast. Bones of the prosauropod *Sarahsaurus* have been found in the slightly younger Kayenta Formation in Arizona. In addition, bones of the prosauropod *Seitaad* as well as many prosauropod tracks have been found in the Navajo Sandstone Formation, so these dinosaurs were also certainly around at the time the Moenave Formation was being deposited.

Similarly, ornithischians other than *Scutellosaurus* existed elsewhere in the Early Jurassic world. In particular, a group of ornithischians called heterodontosaurs are well known in the Early Jurassic of Africa and South America, and an undescribed heterodontosaur is known from the Kayenta Formation of Arizona. Heterodontosaurs are also good candidates for the makers of at least some *Anomoepus* tracks. In England, bones of a large, quadrupedal, armored relative of *Scutellosaurus*, called *Scelidosaurus* (see "Highlight: The Dabney *Scelidosaurus*"), have been found in strata of Early Jurassic age, only slightly younger than the Moenave Formation. *Scelidosaurus* or something similar may well have lived here in Moenave Formation time—new discoveries will tell.

Most "protomammals" became extinct by the end of the Triassic, but a few, such as *Kayentatherium*, survived into the Early Jurassic. True mammals evolved near the end of the Triassic, but in the Early Jurassic, these early mammals were about the size of modern shrews, too small to make the unnamed SGDS tracks. But tiny fossils of Early Jurassic mammals are known from many other places in the world, so they likely lived around Lake Dixie, too, and would have left very tiny footprints. Both "protomammals" and mammals would have taken advantage of the Lake Dixie water source.

Because we know that these plants and animals must have lived in the area during Lake Dixie time, it is fairly safe to "paint" them into our reconstruction of Lake Dixie as well, even though we do not yet have any evidence of them in the Moenave Formation. More importantly, we can predict that with enough searching, fossils of these, and probably of other organisms, too, will eventually be found here. So when you are hiking or playing outdoors in southwestern Utah, keep a keen eye on the rocks around you—you may just be the next person to discover an important piece of the Moenave "snapshot." But always be mindful of the laws of the land you are on: don't touch or collect vertebrate fossils; report them to the proper authorities instead.

A Lesson of the SGDS

Research and discovery are by no means exhausted at the SGDS; new discoveries will undoubtedly continue to further enhance the diversity of fossils known from the site, as well as its scientific importance. But one of the greatest legacies the SGDS offers is its lesson about scientific ethics and integrity. When Sheldon Johnson made the first amazing discoveries on his property, he could have done any number of things with the fossils: he could have ignored them and destroyed them in order to sell his property more quickly to a developer, or he could have sold them all. Instead, he and his wife, LaVerna, recognized that the fossils had scientific and educational value, and that this

value was far greater than any amount of money. They set about ensuring that these remarkable treasures would be preserved, protected, and made available to the public for educational purposes, as well as to scientists. Fossils are often bought and sold, and sometimes specimens that would be very important for increasing our understanding of ancient life are never seen by scientists or anyone else interested in geology and paleontology. The Johnsons' remarkable generosity means that the fossils at the SGDS will continue to teach and be enjoyed by generations of people far into the future. Furthermore, some of the fossils found at the SGDS have been named after the people who discovered them or who helped preserved them at the museum. If you discover any fossils while in the St. George area, *leave them where they are*, but tell the SGDS and Dixie State University paleontologists—if the fossils belong to something new, it could be named after you!

Highlight: The Dabney *Scelidosaurus*

Although no fossils—either tracks or bones—of the armored dinosaur *Scelidosaurus harrisonii* have ever been found in North America, the SGDS museum has on display a beautiful replica of the finest specimen ever found of this unusual dinosaur, which David Sole discovered in 2000. The only place in the world where fossils of this dinosaur have been discovered is southwestern England (a few fragments from Arizona, China, and South Africa have been attributed to *Scelidosaurus* in the past, but few paleontologists today agree with those assignments). The replica at the SGDS (SGDS 1311) was obtained from a generous donation to the museum by Mr. Virginius "Jinks" Dabney and his wife, Barbara, in 2010. This replica is the only example of *Scelidosaurus* in any museum outside England or Ireland. *Scelidosaurus* and its relatives may have made tracks

A

called *Moyenisauropus*, which have not been found in St. George or in the Moenave Formation, but have been found in Early Jurassic–age strata in Africa and Europe. Some tracks from the Navajo Sandstone Formation in Arizona may be *Moyenisauropus* tracks.

Scelidosaurus comes from rocks that are only a few million years younger than the fossil-bearing rocks at the SGDS. It is quite possible that something very similar to *Scelidosaurus* lived during the time and in the area of Lake Dixie and is waiting to be discovered. Work is in progress to obtain more skeletal replicas of Early Jurassic dinosaurs and other animals for display in the SGDS museum alongside the tracks, and donations for that purpose are always welcome and greatly appreciated.

5.2. *A,* Replica of the best-known skeleton of the early armored ornithischian dinosaur *Scelidosaurus* on display at the SGDS (SGDS 1311). *B,* The early armored dinosaur *Scelidosaurus.* *(Restoration by H. Kyoht Luterman.)*

B

GLOSSARY

Anchisauripus (AYNG-kee-SAWR-ih-puhs)—an **ichnogenus** made by medium-sized **theropod** dinosaurs. The name means "foot of *Anchisaurus*," a dinosaur that was once thought to have made this kind of footprint.

Anomoepus (AN-oh-MEE-puhs)—an **ichnogenus** made by small **ornithischian** dinosaurs. Its toe impressions have different proportions than those of *Grallator*, and they are arranged in a different way. The name means "dissimilar foot."

Araucarites (AR-ow-kah-RY-teez)—an extinct **genus** of **conifer**. The name means "similar to *Araucaria*" (a living type of conifer).

basin—a low-lying area in which eroded sediment from upland areas can accumulate.

Batrachopus (BAT-ray-KOH-puhs)—an **ichnogenus** made by small, early relatives of modern crocodylians. The name means "frog foot."

bennettitalean (BEH-neh-tih-TAY-lee-ihn)—any member of a group of now-extinct plants that was common in the Mesozoic Era. Some looked vaguely like skinny palm trees; others looked like squat pineapples on the ground with vaguely flower-like reproductive structures on the trunk.

Bifurculapes (BIH-fur-KYOO-lah-pehs)—an **ichnogenus** made by beetles. The name means "two-forked foot."

bipedal (by-PEE-duhl)—moving around on two limbs. Animals that move this way are called bipeds.

body fossil—any fossil that was at one time part of a living organism's body. Examples are fossil bones, teeth, claws, shells, scales, skin, leaves, cones, and wood. Compare with **trace fossil**.

Brasilichnium (BRA-zihl-IHK-nee-um)—an **ichnogenus** made by early "**protomammals**." The name means "track from Brazil," where the track type was first discovered.

carnivore—an organism that eats meat.

Ceratodus (SEH-ruh-TOH-dus)—a **genus** of **lungfish** that lived from the Late **Triassic** into the **Cretaceous**. The name means "horned tooth."

Cenozoic Era (SEH-noh-ZOH-ihk)—the division of the **Phanerozoic Eon** from 66 million years ago to today.

Characichnos (KAHR-ak-IHK-nohs)—an **ichnogenus** made by swimming dinosaurs and other land animals. The name means "sharp scrape trace."

chevron marks—a series of V-shaped marks made by small objects pushed through loose sediment by water currents.

Chinle Formation (CHIN-lee)—a set of **strata** that was deposited in the American Southwest during the Late **Triassic**, between 225 and 201 million years ago.

Chinlea (chin-LEE-uh)—a **genus** of **coelacanth** that lived during the Late **Triassic**. The name means "from the Chinle Formation."

Clathropteris (klath-RAHP-ter-is)—an extinct **genus** of **fern**. The name means "lattice fern."

Clevosaurus (KLEE-voh-SAWR-uhs)—a **genus** of **sphenodontian** that lived during the Late **Triassic** and Early **Jurassic**. The name means "Clevum lizard," after where it was first discovered, in Gloucestershire, England.

coelacanth (SEEL-uh-kanth)—any member of a group of fishes characterized by thick, heavy limb bones and a small central lobe-fin on the tail. Coelacanths are fairly close relatives of the first **vertebrates** ever to walk on land. The name means "hollow spine."

Coelophysis (SEE-loh-FY-sihs)—a **genus** of **coelophysoid** dinosaur that lived during the Late **Triassic**. The name means "hollow form."

coelophysoid (SEE-loh-FY-soyd)—any member of a group of early, primitive **theropod** dinosaurs that lived during the Late **Triassic** and Early **Jurassic**. Coelophysoids were skinny, long necked, and fast moving. The name means "hollow form–like."

conchostracan (kahn-KAH-strih-kihn)—any member of a group of tiny **crustaceans** that lives inside two clam-like shells in lakes and ponds. Sometimes also called "clam shrimp," though it is not a shrimp at all.

conglomerate—**sedimentary rock** made up of rounded grains of many different sizes, including anything bigger than sand-sized grains (0.079 in. [2 mm]) down to microscopic, clay-sized particles (smaller than 0.00015 in. [0.004 mm]).

conifer—any member of a group of trees that bears its seeds inside woody cones. Modern examples include pines, firs, junipers, spruces, and redwoods.

correlation—the process of determining whether **strata** in different places are the same age by assessing whether they contain identical fossils. If they do, then the strata **correlate** with one another.

Cretaceous Period (kreh-TAY-shus)—the last division of the **Mesozoic Era**, from 145 to 66 million years ago. The Cretaceous is further divided into the Early Cretaceous (145–100.5 million years ago) and Late Cretaceous (100.5–66 million years ago).

crustacean (kruh-STAY-shihn)—any member an enormous group of many-legged animals that includes barnacles, crabs, lobsters, shrimp, and pillbugs, as well as many less familiar animals, such as **ostracods**, **conchostracans**, **notostracans**, and many extinct forms.

cyanobacteria (sy-AN-oh-bak-TEER-ee-uh)—bacteria that can photosynthesize food the same way plants do. Cyanobacteria often build **stromatolites**.

cycad (SY-kad)—any member of a particular group of plants distantly related to **conifers**. Like conifers, cycads bear their seeds in long, woody cones. A few cycads still live in today's tropics, but they were more common during the **Mesozoic Era**.

daughter element—an element whose atoms form as the result of the decay of previously unstable atoms of a **parent element**.

digitigrade (dih-JIH-tih-grayd)—moving with only the fingers and/or toes in contact with the ground, but not the **metatarsus**.

Dilophosaurus (dy-LOH-foh-SAWR-uhs)—a **genus** of **theropod** dinosaur that lived during the Early **Jurassic**. The name means "two-crested lizard."

Dimorphodon (dy-MOHR-foh-dahn)—a **genus** of **pterosaur** that lived during the Early **Jurassic**. The name means "two-shaped teeth," for the different kinds of teeth in its mouth.

Dinosaur Canyon Member—the lower division of the **Moenave Formation**. Sediments in the Dinosaur Canyon Member were deposited by streams in channels and floodplains at the very end of the **Triassic Period** and the very beginning of the **Jurassic Period**.

Diplichnites (DIH-plihk-NY-teez)—an **ichnogenus** made by an arthropod. The name means "twofold trace."

eon—the largest division of the geologic time scale. The time scale includes two eons: the **Precambrian** and the **Phanerozoic**.

epoch (EH-puhk)—a division of a **period** on the geologic time scale.

Equisetum (EH-kwih-SEE-tuhm)—a **genus** of **horsetail**. The name means "horse bristle."

era—a division of an **eon** on the geologic time scale. In the **Phanerozoic Eon**, there are three eras: the **Paleozoic**, **Mesozoic**, and **Cenozoic**.

erg—a vast area of land covered mostly by sand dunes.

Eubrontes (yoo-BRAHN-teez)—an **ichnogenus** made by large **theropod** dinosaurs. The name means "true thunder," presumably for the fanciful sound the track makers made when walking.

Exocampe (EKS-oh-KAM-pee)—an **ichnogenus** possibly made by early **sphenodontians**. The name means "bent outward," describing the arrangement of some of the toe impressions.

facultative—optional; by choice. Facultative **bipeds** (or facultative **quadrupeds**) can move either on just their hind limbs or on all four limbs. The opposite of facultative is **obligate**.

fern—any member of a group of plants with large, lacy leaves that reproduce with spores instead of seeds. Ferns typically live in warm, wet places.

flute mark—a very large groove carved into sediment by strong currents. Flute marks are a **sedimentary structure** and are oriented in the same direction as the currents that made them.

formation—a package, usually a layer, of rock found over a wide area that is different from others above and below it.

fossil—any remains or traces of ancient life. Very few fossils are of dinosaurs.

genus (plural: **genera**) (JEE-nuhs; plural: JEH-neh-ruh)—a group of closely related organisms. Genera contain smaller groupings of even more closely related organisms called **species**.

geologist—a scientist who studies the Earth, usually by examining rocks and their components.

geology—the study of the Earth, usually focusing on rocks and their components.

Gigandipus (JY-gan-DIH-puhs)—an **ichnogenus** made by large **theropod** dinosaurs. It differs from *Eubrontes* in having an impression of the small first toe. The name means "giant two foot," for the sizes of the track makers.

ginkgo—any member of a group of plants with small, heart-shaped leaves and seeds in big, fleshy pods. Today, there is only one kind of ginkgo (called the maidenhair tree), but in the **Mesozoic Era**, several different kinds were fairly common.

Grallator (GRAW-lay-tohr)—an **ichnogenus** made by small **theropod** dinosaurs. The name means "stilt walker" because the track makers were originally thought to be long-legged birds.

Helminthoidichnites (hehl-MIHN-thoyd-ihk-NY-teez)—an **ichnogenus** of horizontal burrow made by **invertebrates**, probably worms. The name means "worm-like trace."

herbivore—an organism that eats plants.

heterodontosaur (HEH-ter-oh-DAHN-toh-sawr)—any member of a group of early **bipedal, herbivorous, ornithischian** dinosaurs characterized by short, tusk-like teeth near the front of the jaws. The name means "different kinds of tooth lizard," for the different-shaped teeth throughout their mouths.

Hettangian Stage (heh-TAN-jee-ihn)—the earliest division of the **Jurassic Period**, from 201.3 to 199.3 million years ago. Named for the town of Hettange-Grande in France.

horsetail—a plant with long, needle-like leaves attached in a circular pattern on the rough stem at intervals. Only one type of horsetail survives today, and it prefers warm, wet environments. In the **Paleozoic** and **Mesozoic Eras**, there were other kinds of horsetails, some as large as trees.

hybodont (HY-boh-dahnt)—any member of a group of now-extinct sharks characterized by bony spines in front of the dorsal fins on the back and small, blunt teeth for crushing shelled **invertebrates**. Many hybodonts lived in freshwater. The name means "hump tooth."

ichnite (IHK-nyt)—a **trace fossil**.

ichnogenus (plural: **ichnogenera**) (IHK-noh-JEE-nuhs; plural: IHK-noh-JEH-neh-ruh)—a group of very similar **trace fossils**. Ichnogenera contain **ichnospecies**.

ichnology (ihk-NAH-loh-jee)—the study of **trace fossils**.

ichnospecies (IHK-noh-SPEE-sees)—any member of a type of **trace fossil** with features distinguishing it from other, different ichnospecies. Similar ichnospecies are grouped together in **ichnogenera**.

igneous rock—any rock that forms from solidified **magma**. There are two kinds of igneous rock: intrusive (or plutonic), which forms when magma solidifies underground, and extrusive (or volcanic), which forms when magma solidifies at the surface (magma at the surface is called **lava**).

invertebrate (ihn-VER-tih-briht)—any animal that does not have mineralized or cartilaginous backbones or bones of any kind. Some invertebrates have a structure like a skeleton, but this is not made of bone or cartilage. Animals that have skeletons made of bone or cartilage are **vertebrates**.

joint—a natural fracture in a body of rock. A joint is similar to a fault, except that in a fault, one or both sides of the rock body have moved along the fracture, whereas in a joint, no such movement has occurred.

Jurassic Period (jur-ASS-ihk)—the middle division of the **Mesozoic Era**, from 201 to 145 million years ago. The Jurassic is further divided into the Early Jurassic (201–174.1 million years ago), Middle Jurassic (174.1–163.5 million years ago), and Late Jurassic (163.5–145 million years ago).

Kayenta Formation—a set of **strata** that was deposited by streams in channels, ponds, and floodplains in the American Southwest during the Early **Jurassic**, between about 198 and 195 million years ago.

Kayentapus (kah-YEHN-tuh-puhs)—an **ichnogenus** made by large **theropod** dinosaurs. It differs from *Eubrontes* in having longer outer toes. The name means "foot from the **Kayenta Formation**."

Kayentatherium (kah-YEHN-tuh-THEER-ee-uhm)—a **genus** of **"protomammal"** that lived during the Early **Jurassic**. The name means "Kayenta Formation beast."

Kayentavenator (kah-YEHN-tuh-VEH-nah-tohr)—a **genus** of **theropod** dinosaur that lived during the Early **Jurassic**. The name means "Kayenta Formation hunter."

Kouphichnium (koo-FIHK-nee-uhm)—an **ichnogenus** made by horseshoe crabs. The name means "lightweight trace."

Lake Dixie—an ancient lake that covered parts of southwestern Utah, northwestern Arizona, and southern Nevada at the beginning of the **Jurassic Period**. Sediments deposited in this lake are now preserved as the **Whitmore Point Member** of the **Moenave Formation**. To the east and north of Lake Dixie during the Early Jurassic was a giant **erg** that is now preserved as the **Wingate Sandstone Formation**.

lava—liquid (melted) rock that emerges from deep underground and solidifies at the surface to form certain kinds of **igneous rock**.

Lissodus (lih-SOH-dus)—a **genus** of **hybodont** shark that lived from the beginning of the **Triassic** into the **Cretaceous**. The name means "smooth tooth."

lithify (LITH-ih-fy)—to turn loose sediment into sedimentary rock. Lithification typically happens when heavy, overlying sediment compacts sedimentary grains together and/or when groundwater deposits minerals in the tiny spaces between grains, cementing them together.

Lockeia (LAH-kee-uh)—an **ichnogenus** made by small clams and possibly **conchostracans**. Named for nineteenth-century paleontologist John Locke.

Lophionotus (LOH-fee-oh-NOH-tuhs)—a **genus** of **semionotid** fish that lived during the Late **Triassic** and Early **Jurassic**. The name means "ridge back."

lungfish—any member of a group of large, vaguely eel-like, freshwater fishes with thick, heavy limb bones. As their name implies, lungfish have lungs and can breathe air. Only three kinds of lungfish are alive today, but now-extinct kinds were fairly common in the **Paleozoic** and **Mesozoic Eras**.

magma—rock in liquid form. Most magma is deep underground; magma that breaks through to the surface, such as from a volcano, is called **lava**.

Megapnosaurus (meh-GAP-noh-SAWR-uhs)—a **genus** of **coelophysoid** dinosaur that lived during the Early **Jurassic**. The name means "large, dead lizard." *Megapnosaurus* was once called *Syntarsus*, but that name had already been given to a beetle from Madagascar, so the dinosaur's name had to be changed.

member—a subdivision of a **formation**.

Mesozoic Era (MEH-soh-ZOH-ihk)—the division of the **Phanerozoic Eon** from 253 to 66 million years ago. The Mesozoic is divided into the **Triassic**, **Jurassic**, and **Cretaceous** Periods.

metamorphic rock—rock that forms when preexisting rock is subjected to either high temperature or high temperature and pressure, which alters its chemical structure.

metatarsus (MEH-tuh-TAR-suhs)—the part of the (hind) leg between the ankle and the base of the toes. Bones in the metatarsus are called metatarsals. In some animals, the metatarsals fuse together into a single bone.

Milnerites (MIHL-ner-I-teez)—an extinct **genus** of **conifer**. Named for SGDS paleontologist Andrew R. C. Milner, who discovered the fossils.

Moenave Formation (MOH-eh-NAH-vee)—a set of **strata** that was deposited in the American Southwest during the Early **Jurassic**, between about 201 and 198 million years ago. The Moenave Formation is divided into a lower **Dinosaur Canyon Member**, which was deposited by streams in channels and floodplains, and an upper **Whitmore Point Member**, which was deposited in the huge **Lake Dixie**.

Moenkopi Formation (MOH-ehn-KOH-pee)—a set of **strata** that was deposited in the American Southwest in shallow seas, beaches, and tidal flats during the Early **Triassic**, between about 250 and 245 million years ago. In Arizona and New Mexico, the age of the Moenkopi Formation extends into the early part of the Middle **Triassic**.

Moyenisauropus (moy-EH-nih-sawr-OH-puhs)—an **ichnogenus** made by large **ornithischian** dinosaurs. The name means "Moyeni lizard foot" (Moyeni is a small town in South Africa).

mud cracks—**sedimentary structures** created naturally when previously wet mud dries out and contracts, cracking the surface into irregular shapes. Also called desiccation cracks.

mudstone—**sedimentary rock** made of grains smaller than 0.00015 in. (0.0038 mm). Mudstones do not have the paper-thin, internal layering of **shales** and break apart into irregular chunks.

natural cast—a naturally occurring replica of a **trace fossil**, such as a footprint, that forms when sediment fills in an actual trace (**natural mold**) and then **lithifies**.

natural mold—a negative impression in sediment created by a body part of an organism—for example, a footprint. If preserved as a fossil, a natural mold is a form of **trace fossil**. A natural mold may be associated with a **natural cast**.

Navajo Sandstone Formation (NAH-vuh-hoh)—a set of thick **strata** that was deposited in an enormous **erg** at the end of the Early **Jurassic**. The Navajo erg extended from southern California to central Colorado and New Mexico, covering much of Utah and parts of southern Idaho and Wyoming.

notostracan—(NOH-toh-STRAY-kihn)any member of a group of tiny **crustaceans** that looks vaguely like a two-tailed tadpole and lives in shallow lakes and ponds. Sometimes also called "tadpole shrimp" or "shield shrimp," though it is not a shrimp at all.

obligate—mandatory; required. Obligate bipeds move on their hind limbs only; obligate quadrupeds move on all four limbs. The opposite of obligate is **facultative**.

ornithischian (OR-nih-THIH-shee-uhn)—any member of the **herbivorous**, "bird-hipped" group of dinosaurs. The other dinosaur group is the saurischians, or "lizard-hipped" dinosaurs.

ostracod (AH-struh-kahd)—any member of a group of tiny **crustaceans** that lives inside two clam-like shells in lakes and ponds and in oceans. Sometimes also called "seed shrimp," though it is not a shrimp at all.

overprinting—the process by which a **quadrupedal** animal, when walking, puts its hind foot down in the same spot that the forefoot was just in, leaving a footprint over part or all of the handprint.

Pagiophyllum (PAJ-ee-oh-FY-luhm)—an extinct **genus** of **conifer**. The name means "solid leaf."

palaeoniscoid (PAY-lee-oh-NIH-skoyd)—any member of a group of now-extinct fishes characterized by slender bodies; thick, heavy bones; and thick, heavy scales. The name means "old codfish-like form."

paleontologist (PAY-lee-ahn-TAH-loh-jihst)—a scientist who studies ancient life, usually by examining fossils.

paleontology (PAY-lee-ahn-TAH-loh-jee)—the study of ancient life, usually fossils, including not only dinosaurs but any kind of formerly living thing, such as other animals, plants, and even ancient bacteria.

Paleophycus (PAY-lee-oh-FY-kus)—an **ichnogenus** of horizontal burrow made by **invertebrates**, probably worms. The name means "old seaweed" because the fossils were once thought to be of sea plants or algae.

Paleozoic Era (PAY-lee-oh-ZOH-ihk)—the division of the **Phanerozoic Eon** from 541 million to 253 million years ago.

Pangaea (pan-JEE-uh)—the last giant supercontinent (a single land mass made up of all of today's separate continents joined together) on Earth. Pangaea slowly began to form more than half a billion years ago but was not completely assembled until about 300 million years ago. It remained a supercontinent until about 200 million years ago, when it broke apart into two smaller supercontinents called Laurasia (a northern landmass made up of today's North America, Europe, and most of Asia) and Gondwana (a southern landmass made up of today's South America, Africa, India, Australia, and Antarctica). The name means "all Earth."

Panthalassa (PAN-thuh-LASS-uh)—a single, gigantic ocean that surrounded the supercontinent of **Pangaea**. Panthalassa divided into several smaller oceans as Pangaea broke apart, beginning about 200 million years ago.

parent element—an element made up of unstable atoms that have not yet undergone radioactive decay, which will turn them into atoms of a different, **daughter element**.

Parundichna (PEHR-uhn-DIHK-nuh)—an **ichnogenus** made by swimming fish. The name means "near undulating (wavy) trace."

pentadactyl (PEHN-tuh-DAK-tuhl)—having five digits (fingers or toes) on the hand or foot.

period—a division of an **era** on the geologic time scale.

Petrified Forest Member—the upper division of the **Chinle Formation** in southwestern Utah. Sediments in the Petrified Forest Member, including large amounts of volcanic ash, were deposited by streams in channels and floodplains in the Late Triassic. The name comes from Petrified Forest National Park in Arizona, where these rocks contain abundant petrified wood.

Phanerozoic Eon (FAN-er-oh-ZOH-ihk)—the most recent approximately one-eighth of Earth's history, from 541 million years ago to the present. The name means "visible animal life."

piscivore (PY-sih-vohr)—an organism that eats fish.

plantigrade (PLAN-tih-grayd)—moving with not only the fingers and/or toes but also the bones between the wrist and base of the fingers (the metacarpus) and/or bones between the ankle and base of the toes (the **metatarsus**) in contact with the ground.

Podozamites (POH-doh-za-MY-teez)—an extinct **genus** of **conifer**. The name means "*Zamia* foot form" (*Zamia* is a type of living plant.)

Precambrian Eon (pree-KAM-bree-ihn)—the first approximately seven-eighths of Earth's history, from the formation of the Earth, 4.6 billion years ago, until 541 million years ago. The name means "before the Cambrian (era)."

prosauropod (pro-SAWR-oh-pahd)—any member of a group of **herbivorous**, **bipedal** dinosaurs characterized by a very long neck and small head.

Prosauropods were early relatives of sauropods, which include the largest land animals that ever lived; well-known sauropods include *Apatosaurus*, *Diplodocus*, and *Brachiosaurus*. The name means "before reptile foot."

"protomammal"—any member of a group of now-extinct animals that includes the ancestors of mammals.

Protosuchus (PROH-toh-SOO-kuhs)—a **genus** of early crocodylian that lived during the Early **Jurassic**. The name means "first crocodile."

Protovirgularia (PROH-toh-VUR-gyoo-LEH-ree-uh)—an **ichnogenus** made by **scaphopods** or by dragonfly larvae. The name means "first twig form."

pterosaur (TEH-roh-sawr)—any member of a now-extinct group of reptiles characterized by the ability to fly using wings made of stiffened skin supported by a very long fourth finger on the hand. Pterosaurs are not dinosaurs, though they are close relatives. Often incorrectly called pterodactyls, pterosaurs lived alongside dinosaurs from the Late **Triassic** through the end of the **Cretaceous**. The name means "wing lizard."

quadrupedal (KWAH-droo-PEE-duhl)—moving around on four limbs. Animals that move this way are called quadrupeds.

Quayia (KWAY-ee-uh)—a **genus** of **coelacanth** that lived during the Late **Triassic**. The name means "from Quay County" (New Mexico).

radioactive decay—the process by which an unstable atom spontaneously tries to become stable by either ejecting particles from its nucleus or absorbing other particles into its nucleus.

raindrop impressions—**sedimentary structures** that form when drops of rain hit loose sediment, creating small craters.

Rhamphinion (ram-FIH-nee-uhn)—a **genus** of **pterosaur** that lived during the Early **Jurassic**. The name means "beak nape."

ridge—in terms of a specialized kind of **sedimentary structure**, an elevated, linear area between **swales**.

rill marks—**sedimentary structures** that form when trickles of water run down a slope, carving shallow, frilly grooves in loose sediment.

ripple marks—a series of long and straight or short and horseshoe-shaped **sedimentary structures** that form when water currents or wind pushes loose sediment into mounds. Ripple marks are excellent indicators of the direction in which water currents flowed or winds blew.

Saintgeorgia—an extinct **genus** of **conifer**. The name means "from St. George."

sandstone—**sedimentary rock** made of grains between 0.0025 and 0.0790 in. (0.06–2.00 mm).

Sanmiguelia (SAN-mihg-EL-ee-uh)—an extinct **genus** of plant whose relationships are unknown. The name means "from (the) San Miguel (River)."

Sarahsaurus (SEH-ruh-SAWR-uhs)—a **genus** of **prosauropod** dinosaur that lived during the Early **Jurassic**. The name means "Sarah's lizard" (Sarah Butler was the discoverer of the fossils).

scale scratch lines—**trace fossils** created when scales on an animal's skin carve grooves in loose sediment.

scaphopod (SKAY-foh-pahd)—any member of a group of ocean-living invertebrates sometimes called a "tusk shells" or "cleft-footed clams."

Scelidosaurus (skeh-LIH-doh-SAWR-uhs)—a **genus** of armored **ornithischian** dinosaur that lived during the Early **Jurassic**. The name means "leg lizard."

Scoyenia—an **ichnogenus** of horizontal burrow made by **invertebrates**, possibly insect larvae. Named for former Grand Canyon chief ranger E. T. Scoyen.

Scutellosaurus (skoo-TEHL-oh-SAWR-uhs)—a **genus** of armored **ornithischian** dinosaur that lived during the Early **Jurassic**. The name means "lizard with small shields."

sedimentary rock—any rock that is made either of pieces of preexisting rock that have been weathered into bits (called clastic or detrital sedimentary rock, such as **shale**, **mudstone**, **sandstone**, **siltstone**, or **conglomerate**) or of minerals that have precipitated out of evaporating water (called chemical sedimentary rock, such as limestone, gypsum, or rock salt).

sedimentary structure—any naturally occurring feature created in loose sediment by the action of water or wind. Examples are **mud cracks**, **raindrop impressions**, **ridges** and **swales**, **rill marks**, **ripple marks**, and **tool** and **flute marks**.

seed fern—any member of a group of now-extinct plants that had fern-like leaves but reproduced with seeds instead of spores.

Seitaad (SAY-ee-tawd)—a **genus** of **prosauropod** dinosaur that lived during the Early **Jurassic**. Named for a mythical Navajo monster.

semionotid (SEH-mee-oh-NOH-tihd)—any member of a group of now-extinct fishes characterized by bodies that were tall from top to bottom and short from front to back, and had thick, heavy, diamond-shaped scales. The name, meaning "signal [on the back]," refers to peculiar scales along the back, from the back of the head to the dorsal fin, that bore tiny, backward-pointing spines.

Semionotus (SEH-mee-oh-NOH-tuhs)—a **genus** of **semionotid** fish that lived during the Late **Triassic** and Early **Jurassic**. The name means "ridge back."

SGDS—initials of the St. George Dinosaur Discovery Site at Johnson Farm.

shale—**sedimentary rock** made of grains smaller than 0.00015 in. (0.0038 mm). Shales have paper-thin internal layers called laminae (lamina for just one) and break apart into very thin, smooth chips or sheets.

Shinarump Conglomerate Member (SHIH-nuh-ruhmp)—the lower division of the **Chinle Formation** in southwestern Utah. Sediments in the Shinarump Conglomerate were deposited by large rivers in channels and floodplains in the Late **Triassic**.

siltstone—**sedimentary rock** made of grains between 0.00015 and 0.00250 in. (0.0038–0.0635 mm).

Sinemurian Stage (SIH-neh-MUR-ee-ihn)—the second division of the **Jurassic Period**, from 199.3 to 190.8 million years ago. Named for the town of Semur-en-Auxois in France.

skin impression—a **natural mold** made by an animal's skin in loose sediment and preserved as a fossil. **Natural casts** of skin impressions can also form.

Skolithos (skoh-LIH-thohs)—an **ichnogenus** of vertical burrow made by **invertebrates**, possibly worms or spiders. The name means "stone worm."

species—any member of a group of organisms that can reproduce with other members of the group (of the opposite sex in the case of animals that reproduce sexually). Species (the word is both singular and plural) are the most closely related groups of organisms.

sphenodontian (SFEE-noh-DAHN-tee-ihn)—any member of a group of lizard-like animals related to the tuatara, which is the only living sphenodontian. In the **Mesozoic Era**, especially the Late **Triassic** and **Jurassic**, sphenodontians were very common, much more common than lizards; there were even some swimming species. The name means "wedge tooth."

Springdale Sandstone Member—the lowest division of the **Kayenta Formation** in southwestern Utah. Sediments in the Springdale Sandstone Member were deposited by large rivers in channels and floodplains in the Early **Jurassic**.

stage—a division of an **epoch** on the geologic time scale.

step length—the distance between a right and left footprint in a **trackway**.

stratigraphy (stra-TIHG-rah-fee)—the study of how rock layers (**strata**) form and are ordered. Stratigraphy is a branch of **geology**.

stratum (plural: **strata**) (STRA-tuhm; plural: STRA-tuh)—a layer of sedimentary rock.

stromatolite (stroh-MAT-oh-lyt)—a mound of lithified sediment created over a long period by sediment deposited by water and cemented together by **cyanobacteria**. The name means "mattress form."

swale—a low, valley-like **sedimentary structure** between two **ridges**.

swim track—a **trace fossil** made by an animal while it is swimming. Typical swim tracks are long grooves created when the claws at the ends of the digits of a swimming animal are able to strike bottom and carve through loose sediment.

tetradactyl (TEH-truh-DAK-tuhl)—having four digits (fingers or toes) on the hand or foot.

theropod (THEH-roh-pahd)—any member of a group of dinosaurs characterized by air-filled bones, wishbones, and feet that left **tridactyl** footprints, among other features. Most theropods were **carnivores**, though a few were **herbivores**. Birds evolved from earlier theropod dinosaurs and are therefore a kind of theropod—they are the only living dinosaurs. The name means "beast foot."

tool mark—a groove carved into sediment when an object such as a twig or rock is pushed by water currents. Tool marks are a kind of **sedimentary structure**.

trace fossil—any fossil that was made by a part of a living organism's body but is *not* itself a part of that organism's body. Examples are footprints, coprolites (fossil feces), and bite marks made on a leaf, shell, or bone. Compare with **body fossil**.

trackway—a series of tracks made one after another by the same animal as it moves.

Triassic Period (try-ASS-ihk)—the first division of the **Mesozoic Era**, from 253 to 201.3 million years ago. The Triassic is itself divided into the Early Triassic (253–247.2 million years ago), Middle Triassic (247.2–237 million years ago), and Late Triassic (237–201.3 million years ago).

tridactyl (try-DAK-tuhl)—having three digits (fingers or toes) on the hand or foot.

undertrack—an imperfect, distorted version of a footprint that forms in layers of loose sediment beneath the one into which an animal actually steps.

Undichna (uhn-DIHK-nuh)—an **ichnogenus** made by swimming fish. The name means "undulating (wavy) trace."

unguligrade (uhn-GYOO-lih-grayd)—moving with only the very tips of the fingers and/or toes in contact with the ground.

vertebra (plural: **vertebrae**)—any of the bones that run down the midline of a **vertebrate**'s body from the base of the head through the tail. Sometimes also called a "backbone."

vertebrate (VER-tih-briht)—any animal that has a skeleton, including **vertebrae**, made either of bone or cartilage, and whose ancestors had such a skeleton.

Wingate Sandstone Formation—a thick set of **strata** that was deposited in an enormous **erg** at the end of the Late **Triassic** and into the Early **Jurassic**. The Wingate erg covered parts of what are now western Colorado, eastern and central Utah, northern New Mexico, and northern Arizona. The Wingate erg existed at the same time sediments were being deposited in **Lake Dixie**.

Whitmore Point Member—the upper division of the **Moenave Formation**. Sediments in the Whitmore Point Member were deposited in and around **Lake Dixie** at the beginning of the **Jurassic period**.

SUGGESTED

FURTHER READING

Much more information is available in several other books and in papers in scientific journals. For more information specifically about the SGDS and its fossils, there are many papers in the following volume (which will be referred to in subsequent listings as Harris et al., "Triassic-Jurassic Terrestrial Transition"):

Harris, J. D., S. G. Lucas, J. A. Spielmann, M. G. Lockley, A. R. C. Milner, and J. I. Kirkland, eds. 2006. "The Triassic-Jurassic Terrestrial Transition." *New Mexico Museum of Natural History and Science Bulletin* 37:1–605. http://econtent.unm.edu/cdm/compoundobject/collection/bulletins/id/265/rec/38.

For further information on the SGDS, we recommend the following (listed from most general to most technical):

Kirkland, J. I., M. G. Lockley, and A. R. C. Milner. 2002. "The St. George Dinosaur Tracksite." *Utah Geological Survey Notes* 34(3): 4–5, 12. http://geology.utah.gov/surveynotes/articles/pdf/stgeorge_tracksite_34-3.pdf.

Milner, A. R. C., and M. G. Lockley. 2006. "History, Geology and Paleontology: St. George Dinosaur Discovery Site at Johnson Farm, Utah." In *Making Tracks across the Southwest: The 2006 Desert Symposium Field Guide and Abstracts from Proceedings*, edited by R. E. Reynolds, 35–48. California State University, Desert Studies Consortium, and LSA Associates, Inc. http://nsm.fullerton.edu/dsc/images/DSCdocs/2006makingtracks.pdf.

Milner, A. R. C., J. D. Harris, M. G. Lockley, J. I. Kirkland, and N. A. Matthews. 2009. "Bird-Like Anatomy, Posture, and Behavior Revealed by an Early Jurassic Theropod Dinosaur Resting Trace." *PLoS ONE* 4(3): e4591. http://dx.doi.org/10.1371/journal.pone.0004591.

Lockley, M. G., J. I. Kirkland, and A. R. C. Milner. 2004. "Probable Relationships between the Lower Jurassic Crocodilomorph Trackways *Batrachopus* and *Selenichnus*: Evidence and Implications Based on New Finds from the St. George Area Southwestern Utah." *Ichnos* 11, 143–49.

Further information is available about different kinds of fossils and the geological and paleontological concepts discussed herein, but much of it is more technical, typically in various scientific books and journals that are not readily available at most public libraries. To obtain copies of these, you may have to visit a local college or university library, although many are made freely available online by either the publishers or the authors. If no publisher website is listed below, we recommend searching the Internet for the article title to see whether the author has posted the paper for free download. We recommend the following (listed in each category in alphabetical order by author):

Early Mesozoic Stratigraphy and Geology of the Southwest

Blakey, R. C., and W. Ranney. 2008. *Ancient Landscapes of the Colorado Plateau.* Grand Canyon, AZ: Grand Canyon Association.

Clemmensen, L. B., H. Olsen, and R. C. Blakey. 1989. "Erg-Margin Deposits in the Lower Jurassic Moenave Formation and Wingate Sandstone, Southern Utah." *Geological Society of America Bulletin* 101(6): 759–73.

Fraser, N. C. 2006. *Dawn of the Dinosaurs: Life in the Triassic.* Bloomington: Indiana University Press.

Heckert, A. B., and S. G. Lucas, eds. 2006. "Upper Triassic Stratigraphy and Paleontology." *New Mexico Museum of Natural History and Science Bulletin* 21. http://econtent.unm.edu/cdm/compoundobject/collection/bulletins/id/919/rec/22.

Kirkland, J. I., and A. R. C. Milner. 2006. "The Moenave Formation at the St. George Dinosaur Discovery Site at Johnson Farm, St. George, Southwestern Utah." In Harris et al., "Triassic-Jurassic Terrestrial Transition," 289–309. http://econtent.unm.edu/cdm/compoundobject/collection/bulletins/id/265/rec/38.

Peterson, F. 1988. "Pennsylvanian to Jurassic Eolian Transportation Systems in the Western United States." *Sedimentary Geology* 56(1–4): 207–60.

Sues, H.-D., and N. C. Fraser. 2010. *Triassic Life on Land: The Great Transition.* New York: Columbia University Press.

Tanner, L. H., and S. G. Lucas. 2007. "The Moenave Formation: Sedimentologic and Stratigraphic Context of the Triassic- Jurassic Boundary in the Four Corners Area, Southwestern U.S.A. *Palaeogeography, Palaeoclimatology, Palaeoecology* 244(1–4): 111–25.

Sedimentary Structures

Collinson, J., N. Mountney, and D. Thompson. 2006. *Sedimentary Structures.* 3rd ed. Edinburgh: Dunedin Academic Press.

Lucchi, F. R. 1995. *Sedimentographica: A Photographic Atlas of Sedimentary Structures.* 2nd ed. New York: Columbia University Press.

Trace Fossils—General

Buatois, L. A., and M. G. Mángano. 2011. *Ichnology: Organism-Substrate Interactions in Space and Time.* Cambridge: Cambridge University Press.

Martin, A. J. 2014. *Dinosaurs Without Bones: Dinosaur Lives Revealed by Their Trace Fossils.* New York: Pegasus Books.

Seilacher, A. 2007. *Trace Fossil Analysis.* Berlin: Springer-Verlag.

Dinosaur Tracks

Dalman, S. G., and R. E. Weems. 2012. "A New Look at Morphological Variation in the Ichnogenus *Anomoepus*, with Special Reference to Material from the Lower Jurassic Newark Supergroup: Implications for Ichnotaxonomy and Ichnodiversity." *Bulletin of the Peabody Museum of Natural History* 54(1): 67–124.

Hitchcock, E. 1858. *Ichnology of New England: A Report on the Sandstone of the Connecticut Valley, Especially Its Fossil Footmarks.* Boston: W. White. http://books.google.com/books?id=HvpaAAAAQAAJ&printsec=toc&source=gbs_navlinks_s#v=onepage&q=&f=false.

Lockley, M. 1991. *Tracking Dinosaurs*. Cambridge: Cambridge University Press.

Lockley, M., and J. Peterson. 2002. *A Guide to the Fossil Footprints of the World*. Boulder: Lockley-Peterson Publications.

Lockley, M. G. 2000. "Philosophical Perspectives on Theropod Track Morphology: Blending Qualities and Quantities in the Science of Ichnology." *Gaia Revista de Geociências, Museu Nacional de História Natural (Lisbon)* 15:279–300. http://www. arca.museus.ul.pt/ArcaSite/obj/gaia/MNHNL-0000790-MG-DOC-web.PDF.

Lockley, M. G., and A. P. Hunt. 1995. *Dinosaur Tracks and Other Fossil Footprints of the Western United States*. New York: Columbia University Press.

Lull, R. S. 1953. "Triassic Life of the Connecticut Valley." *Connecticut State Geological and Natural History Survey Bulletin* 81:1–336.

McDonald, N. G. 2010. *Window into a Jurassic World*. Rocky Hill, CT: Friends of Dinosaur State Park and Arboretum.

Milner, A. R. C., T. A. Birthisel, J. I. Kirkland, B. H. Breithaupt, N. A. Matthews, M. G. Lockley, V. L. Santucci, et al. 2012. "Tracking Early Jurassic Dinosaurs across Southwestern Utah and the Triassic-Jurassic Transition." In *Field Trip Guide Book for the 71st Annual Meeting of the Society of Vertebrate Paleontology*, edited by J. W. Bonde and A. R. C. Milner, 1–107. *Nevada State Museum Paleontological Papers* 1.

Milner, A. R. C., M. G. Lockley, and S. B. Johnson. 2006. "The Story of the St. George Dinosaur Discovery Site at Johnson Farm: An Important New Lower Jurassic Dinosaur Tracksite from the Moenave Formation of Southwestern Utah." In Harris et al., "Triassic-Jurassic Terrestrial Transition," 329–45. http://econtent.unm.edu/ cdm/compoundobject/collection/bulletins/id/265/rec/38.

Milner, A. R. C., M. G. Lockley, and J. I. Kirkland. 2006. "A Large Collection of Well-Preserved Theropod Dinosaur Swim Tracks from the Lower Jurassic Moenave Formation, St. George, Utah." In Harris et al., "Triassic-Jurassic Terrestrial Transition," 315–28. http://econtent.unm.edu/cdm/compoundobject/collection/ bulletins/id/265/rec/38.

Olsen, P. E., and E. C. Rainforth. 2003. "The Early Jurassic Ornithischian Dinosaurian Ichnogenus *Anomoepus*." In *The Great Rift Valleys of Pangea in Eastern North America*. Vol. 2, *Sedimentology, Stratigraphy, and Paleontology*, edited by P. M. Letourneau and P. E. Olsen, 314–68. New York: Columbia University Press.

Olsen, P. E., J. B. Smith, and N. G. McDonald. 1998. "Type Material of the Type Species of the Classic Theropod Footprint Genera *Eubrontes*, *Anchisauripus* and *Grallator* (Early Jurassic, Hartford and Deerfield Basins, Connecticut and Massachusetts, USA)." *Journal of Vertebrate Paleontology* 18(3): 586–601.

Welles, S. P. 1971. "Dinosaur Footprints from the Kayenta Formation of Northern Arizona." *Plateau* 44:27–38.

Whyte, M. A., and M. Romano. 2001. "A Dinosaur Ichnocoenosis from the Middle Jurassic of Yorkshire, UK." *Ichnos* 8(3–4): 223–34.

Williams, J. A. J., A. R. C. Milner, and M. G. Lockley. 2006. "A New Early Jurassic (Hettangian) Track-Bearing Horizon from the Moenave Formation, St. George Dinosaur Discovery Site at Johnson Farm, Washington County, Utah." In Harris et al., "Triassic-Jurassic Terrestrial Transition," 346–51. http://econtent.unm.edu/cdm/ compoundobject/collection/bulletins/id/265/rec/38.

Batrachopus

Lockley, M. G., and C. Meyer. 2004. "Crocodylomorph Trackways from the Jurassic to Early Cretaceous of North America and Europe: Implications for Ichnotaxonomy." *Ichnos* 11(1–2): 167–78.

Olsen, P. E., and K. Padian. 1986. "Earliest Records of *Batrachopus* from the Southwestern United States, and a Revision of Some Early Mesozoic Crocodilomorph Ichnogenera." In *The Beginning of the Age of Dinosaurs*, edited by K. Padian, 259–73. New York: Cambridge University Press.

Fish Swim Traces

de Gibert, J. M. 2001. "*Undichna gosiutensis*, isp. nov.: A new Fish Trace Fossil from the Jurassic of Utah." *Ichnos* 8(1): 15–22.

Simon, T., H. Hagdorn, M. K. Hagdorn, and A. Seilacher. 2003. "Swimming Trace of a Coelacanth from the Lower Keuper of South-West Germany." *Palaeontology* 46(5): 911–26.

Invertebrate Trace Fossils

Buatois, L. A., and M. G. Mángano. 2004. "Animal-Substrate Interactions in Freshwater Environments: Applications of Ichnology in Facies and Sequence Stratigraphic Analysis of Fluvio-Lacustrine Successions." *Geological Society of London Special Publications* 228:311–33.

Hasiotis, S. T., and T. M. Bown. 1992. "Invertebrate Trace Fossils: The Backbone of Continental Ichnology." In *Trace Fossils*, edited by C. G. Maples and R. R. West, 64–104. *Short Courses in Paleontology* 5. Knoxville: Paleontological Society.

Lucas, S. G., A. J. Lerner, A. R. C. Milner, and M. G. Lockley. 2006. "Lower Jurassic Invertebrate Ichnofossils from a Clastic Lake Margin Facies, Johnson Farm, Southwestern Utah." In Harris et al., "Triassic-Jurassic Terrestrial Transition," 128–36. http://econtent.unm.edu/cdm/compoundobject/collection/bulletins/id/265/rec/38.

Metz, R. 2002. "Nonmarine Cretaceous *Protovirgularia*: Possible Dragonfly Larva Tracemaker." *Ichnos* 9(1–2): 75–76.

Fishes

Elliott, D. K. 1987. "A New Specimen of *Chinlea sorenseni* from the Chinle Formation, Dolores River, Colorado." *Journal of the Arizona-Nevada Academy of Science* 22(1): 47–52.

Gibson, S. Z. 2013. "Biodiversity and Evolutionary History of †*Lophionotus* (Neopterygii: †Semionotiformes) from the Western United States." *Copeia* 2013(4): 582–603.

———. 2013. "A New Hump-Backed Ginglymodian Fish (Neopterygii, Semionotiformes) from the Upper Triassic Chinle Formation of Southeastern Utah." *Journal of Vertebrate Paleontology* 33(5): 1037–50.

Hunt, A. P. 1997. "A New Coelacanth (Osteichthyes: Actinistia) from the Continental Upper Triassic of New Mexico." In *New Mexico's Fossil Record 1*, edited by S. G. Lucas, J. W. Estep, T. H. Williamson, and G. S. Morgan, 25–27. *New Mexico Museum of Natural History and Science Bulletin* 11. http://econtent.unm.edu/cdm/compoundobject/collection/bulletins/id/1049/rec/12.

Long, J. A. 1996. *The Rise of Fishes: 500 Million Years of Evolution*. Baltimore: Johns Hopkins University Press.

Maisey, J. G. 2000. *Discovering Fossil Fishes*. Boulder: Westview Press.

Milner, A. R. C., and J. I. Kirkland. 2006. "Preliminary Review of an Early Jurassic (Hettangian) Freshwater Lake Dixie Fish Fauna in the Whitmore Point Member, Moenave Formation in Southwest Utah." In Harris et al., "Triassic-Jurassic Terrestrial Transition," 510–21. http://econtent.unm.edu/cdm/compoundobject/collection/bulletins/id/265/rec/38.

Milner, A. R. C., J. I. Kirkland, and T. A. Birthisel. 2006. "Late Triassic-Early Jurassic Freshwater Fish Faunas of the Southwestern United States with Emphasis on the Lake Dixie Portion of the Moenave Formation, Southwest Utah." In Harris et al., "Triassic-Jurassic Terrestrial Transition," 522–29. http://econtent.unm.edu/cdm/compoundobject/collection/bulletins/id/265/rec/38.

Olsen, P. E., and A. R. McCune. 1991. "Morphology of the *Semionotus elegans* Species Group from the Early Jurassic Part of the Newark Supergroup of Eastern North America with Comments on the Family Semionotidae (Neopterygii)." *Journal of Vertebrate Paleontology* 11(3): 269–92.

Schaeffer, B. 1967. "Late Triassic Fishes from the Western United States." *Bulletin of the American Museum of Natural History* 135:285–342. http://digitallibrary.amnh.org/dspace/handle/2246/1125.

Schaeffer, B., and D. H. Dunkle. 1950. "A Semionotid Fish from the Chinle Formation, with Consideration of Its Relationships." *American Museum Novitates* 1457:1–29. http://digitallibrary.amnh.org/dspace/handle/2246/2356.

Sphenodontians

Fraser, N. C. 1988. "The Osteology and Relationships of *Clevosaurus* (Reptilia: Sphenodontida)." *Philosophical Transactions of the Royal Society of London B* 321:125–78.

Sues, H.-D., N. H. Shubin, and P. E. Olsen. 1994. "Sphenodontian (Lepidosauria: Rhynchocephalia) from the McCoy Brook Formation (Lower Jurassic) of Nova Scotia, Canada." *Journal of Vertebrate Paleontology* 14(3): 327–40.

Pterosaurs

Padian, K. 1983. "Osteology and Functional Morphology of *Dimorphodon macronyx* (Buckland) (Pterosauria: Rhamphorhynchoidea) Based on New Material in the Yale Peabody Museum." *Postilla* 189:1–44. http://peabody.yale.edu/sites/default/files/documents/scientific-publications/ypmP189_1983.pdf.

———. 1984. "Pterosaur Remains from the Kayenta Formation (?Early Jurassic) of Arizona." *Palaeontology* 27(2): 407–413. http://cdn.palass.org/publications/palaeontology/volume_27/pdf/vol27_part2_pp407-413.pdf.

Unwin, D. 1996. *Pterosaurs: From Deep Time*. New York: Pi Press.

Wellnhofer, P. 1991. *The Illustrated Encyclopedia of Pterosaurs*. New York: Crescent Books.

Witton, M. P. 2013. *Pterosaurs: Natural History, Evolution, Anatomy*. Princeton, NJ: Princeton University Press.

Protosuchus

Colbert, E. H., and C. C. Mook. 1951. "The Ancestral Crocodilian *Protosuchus*." *Bulletin of the American Museum of Natural History* 97:149–82. http://digitallibrary.amnh.org/dspace/handle/2246/413.

Gow, C. E. 2000. "The Skull of *Protosuchus haughtoni*, an Early Jurassic Crocodyliform from Southern Africa." *Journal of Vertebrate Paleontology* 20(1): 49–56.

Theropod Dinosaurs

Colbert, E. H. 1989. "The Triassic Dinosaur *Coelophysis*." *Museum of Northern Arizona Bulletin* 57:1–160.

———. 1995. *The Little Dinosaurs of Ghost Ranch*. New York: Columbia University Press.

Gay, R. 2010. *Notes on Early Mesozoic Theropods*. Raleigh: Lulu Press. http://www. lulu.com/us/en/shop/robert-gay/notes-on-early-mesozoic-theropods/paperback/ product-11028232.html.

Lucas, S. G., and A. B. Heckert. 2001. "Theropod Dinosaurs and the Early Jurassic Age of the Moenave Formation, Arizona-Utah, USA." *Neues Jahrbuch für Geologie und Paläontologie Monatshefte* 2001(7): 435–48.

Milner, A. R. C., and J. I. Kirkland. 2007. "The Case for Fishing Dinosaurs at the St. George Dinosaur Discovery Site at Johnson Farm." *Utah Geological Survey Notes* 39(3): 1–3. http://geology.utah.gov/surveynotes/articles/pdf/fishing_dinos_39-3.pdf.

Rowe, T. 1989. "A New Species of the Theropod Dinosaur *Syntarsus* from the Early Jurassic Kayenta Formation of Arizona." *Journal of Vertebrate Paleontology* 9(2): 125–36.

Welles, S. P. 1984. "*Dilophosaurus wetherilli* (Dinosauria, Theropoda) Osteology and Comparisons." *Palaeontographica Abteilung A* 185(4–6): 85–180.

Prosauropod Dinosaurs

Harris, S. K., A. B. Heckert, S. G. Lucas, and A. P. Hunt. 2002. "The Oldest North American Prosauropod, from the Upper Triassic Tecovas Formation of the Chinle Group (Adamanian: Latest Triassic), West Texas." In *Upper Triassic Stratigraphy and Paleontology*, edited by A. B. Heckert and S. G. Lucas, 249–52. *New Mexico Museum of Natural History and Science Bulletin* 21. http://econtent.unm.edu/cdm/ compoundobject/collection/bulletins/id/919/rec/22.

Lockley, M. G., S. G. Lucas, and A. P. Hunt. 2006. "*Evazoum* and the Renaming of Northern Hemisphere '*Pseudotetrasauropus*': Implications for Tetrapod Ichnotaxonomy at the Triassic-Jurassic Boundary." In Harris et al., "Triassic-Jurassic Terrestrial Transition," 199–206. http://econtent.unm.edu/cdm/ compoundobject/collection/bulletins/id/265/rec/38.

Rowe, T. B., H.-D. Sues, and R. R. Reisz. 2010. "Dispersal and Diversity in the Earliest North American Sauropodomorph Dinosaurs, with Description of a New Taxon." *Proceedings of the Royal Society B: Biological Sciences* 278:1044–53.

Scrtich, J. J. W., and M. A. Loewen. 2010. "A New Basal Sauropodomorph Dinosaur from the Lower Jurassic Navajo Sandstone of Southern Utah." *PLoS ONE* 5(3): 1–17. http://dx.doi.org/ 10.1371/journal.pone.0009789.

Ornithischian Dinosaurs

Colbert, E. H. 1981. "A Primitive Ornithischian Dinosaur from the Kayenta Formation of Arizona." *Museum of Northern Arizona Press Bulletin* 53:1–61.

Norman, D. B. 2001. "*Scelidosaurus*, the Earliest Complete Dinosaur." In *The Armored Dinosaurs*, edited by K. Carpenter, 3–24. Bloomington: Indiana University Press.

Norman, D. B., A. W. Crompton, R. J. Butler, L. B. Porro, and A. J. Charig. 2011. "The Lower Jurassic Ornithischian Dinosaur *Heterodontosaurus tucki* Crompton & Charig, 1962: Cranial Anatomy, Functional Morphology, Taxonomy, and Relationships." *Zoological Journal of the Linnean Society* 163(1): 182–276.

Rosenbaum, J. N., and K. Padian. 2000. "New Material of the Basal Thyreophoran *Scutellosaurus lawleri* from the Kayenta Formation (Lower Jurassic) of Arizona." *PaleoBios* 20(1): 13–23. http://docubase.berkeley.edu/cgi-bin/pl_dochome?query_ src=pl_search&format=pdf&collection=PaleoBios_Archive_Public&id=95&show_ doc=yes.

Santa Luca, A. P. 1980. "The Postcranial Skeleton of *Heterodontosaurus tucki* (Reptilia, Ornithischia) from the Stormberg of South Africa." *Annals of the South African Museum* 79(7): 159–211.

Sereno, P. C. 1991. "*Lesothosaurus*, 'fabrosaurids,' and the Early Evolution of Ornithischia." *Journal of Vertebrate Paleontology* 11(2): 168–97.

"Protomammals"

Kermack, D. 1982. "A New Tritylodont from the Kayenta Formation of Arizona." *Zoological Journal of the Linnean Society* 76(1): 1–17.

Ruta, M., J. Botha-Brink, S. A. Mitchell, and M. J. Benton. 2013. "The Radiation of Cynodonts and the Ground Plan of Mammalian Morphological Diversity." *Proceedings of the Royal Society B: Biological Sciences* 280(1769): 1–10. http://rspb. royalsocietypublishing.org/content/280/1769/20131865.full.

Sues, H.-D., and F. A. Jenkins Jr. 2006. "The Postcranial Skeleton of *Kayentatherium wellesi* from the Lower Jurassic Kayenta Formation of Arizona and the Phylogenetic Significance of Postcranial Features in Tritylodontid Cynodonts." In *Amniote Paleobiology: Perspectives on the Evolution of Mammals, Birds, and Reptiles*, edited by M. T. Carrano, T. J. Gaudin, R. W. Blob, and J. R. Wible, 114–52. Chicago: University of Chicago Press.

Plants

Ash, S. R., A. R. C. Milner, D. Sharrow, and D. Tarailo. 2014. "First Known Post-Triassic Occurrence of the Palm-Like Plant Fossil *Sanmiguelia* Brown." *Utah Geological Association Publication* 43: 511–16.

Tidwell, W. D., and S. R. Ash. 2006. "Preliminary Report on the Early Jurassic Flora from the St. George Dinosaur Discovery Site, Utah." In Harris et al., "Triassic-Jurassic Terrestrial Transition," 414–20. http://econtent.unm.edu/cdm/compoundobject/collection/bulletins/id/265/rec/38

INDEX

Numbers in *italics* refer to illustrations.